John Flint South

Household Surgery

Hints on emergencies

John Flint South

Household Surgery
Hints on emergencies

ISBN/EAN: 9783337403904

Printed in Europe, USA, Canada, Australia, Japan

Cover: Foto ©berggeist007 / pixelio.de

More available books at **www.hansebooks.com**

HOUSEHOLD SURGERY;

OR,

Hints on Emergencies.

C

By JOHN F. SOUTH,

FORMERLY ONE OF THE SURGEONS TO ST. THOMAS'S HOSPITAL.

NEW EDITION, WITH ADDITIONS.

LONDON:

JOHN MURRAY, ALBEMARLE STREET.

1880.

MURRAY'S MODERN COOKERY.

230th Thousand.

MODERN DOMESTIC COOKERY, founded upon Principles of Economy and Practical Knowledge, and adapted for the Use of Private Families. With explanatory Woodcuts. Post 8vo., 5s.

"Unquestionably the most complete guide to the culinary department of domestic economy that has yet been given to the world. Independently of receipts of every description, derived from manuscript collections, and tested by experience, the volume contains instructions, amply illustrated by woodcuts, on every collateral branch of the art of housekeeping."— *John Bull.*

LONDON: PRINTED BY W. CLOWES AND SONS, LIMITED, STAMFORD STREET AND CHARING CROSS.

TO

JOSEPH BERENS, ESQ.,

OF KEVINGTON, KENT,

THIS LITTLE WORK

Is Inscribed

IN ACKNOWLEDGMENT OF MUCH KINDNESS

RECEIVED FROM HIM

BY

THE AUTHOR.

BLACKHEATH PARK, *April*, 1847.

PREFACE

TO THE FOURTH EDITION.

—◆◆◆—

In the days of our great grandmothers, part of
a young lady's education consisted in acquiring
a knowledge of the mysteries of "the still-
room," in which fragrant and simple medical
"waters" were elaborated from wild flowers
and herbs, either for their fragrance, culinary
purposes, or household physic, with which the
ladies-bountiful of the parish supplied the
wants of their poorer neighbours and depen-
dents. Within the last few years ladies have
been led to a higher flight in the healing art,
and, warmly taking to practise some small
surgery, have become diligent pupils at the
so-called Ambulance Lectures, the utility and
fittingness of which are well suited to the
tender care and ability of kind-hearted women.
At these lectures notes have been carefully
taken, practical handiness acquired, and certi-
ficates of efficiency given to the students after

examination, and the results cannot but be highly useful.

During this *furore*, as it really was for a while, I was often asked where my 'Household Surgery; or, Hints for Emergencies,' printed more than forty years ago, might be obtained. It had for very many years a large and extensive sale, and was continually recommended by the ever-to-be-revered Bishop Charles James Blomfield to his rural clergy. On inquiry I found that it has been out of print for some years past, and I have therefore suggested, as it has long ceased to belong to me, that it should be reprinted. Its language and advice are simple and without attempt to teach Anatomy, Physiology, Pathology, and Medicine, by strings of professional words, profitless to the reader, who only wants to know how to treat a burn or scald, or stop a bleeding wound.

JOHN F. SOUTH.

BLACKHEATH PARK,
August 8, 1880.

PREFACE

TO THE THIRD EDITION.

———◆———

It is beyond doubt that " my public," so called
by the biographer of dear Doctor Daniel Dove,.
of Doncaster, still holds in estimation 'Household Surgery,' seeing that within little more than
twelve months a second large edition of the said
libellus, or little book, being exhausted, the
steam-press is again required to meet the public
want of the doctor's temporary substitute, but not
intended supplanter, for reasons stated in the first
edition.

I cannot refrain expressing much gratification,
that this little household manual is by no means
in the alone possession and use of private families,
but that it is far from an uncommon associate
with many more erudite, and some more presuming works on Medicine and Surgery, in the
studios of many of my professional brethren, the
best proof that my motives in writing have been
appreciated by those whose esteem I should
grieve to forfeit, and that the book itself is better
worth than I anticipated under the circumstances
which first put it before the public. Its rapid

sale has also shown that it needs no assistance
beyond "a fair stage and no favour," and that it
is not to be driven out of the market by any
crosspatch or caviller who wishes to make room
for his own cheaper production.

It having been suggested that it would be
advantageous in a little book, now on most folks'
table, to add some hints of what should be done
when poison has been accidentally or purposely
taken, and the prompt determination and coun-
teraction of which is of the first importance, I pre-
vailed on my friend Dr. Gladstone to furnish me
with a short notice of the more common poisons,
and the mode in which they may be managed
forthwith, in the *absence or during the delayed
attendance* of a DOCTOR of either of the three
denominations; and he has executed his task
with much ability and plainness, so that the
most simple person may easily follow out his
directions.

BLACKHEATH PARK,
April, 1851.

PREFACE

TO THE SECOND EDITION.

———•———

A LITTLE more than two years have elapsed and
a large edition of this small volume has been
consumed by "my public," affording proof that
the said public has been, at least, amused, if not
also instructed, as was my original "village
auditory," and I hope I may not be thought very
vain in believing that I have by this household
trifle been of some use in my generation.   Sundry
efforts have been made to palm upon the public
counterfeits of this volume, the only likenesses to
which consisted, in abstracting cautiously some of
its most taking points, and in assuming its name
as closely as possible without exciting the
displeasure of the law.

As the object of this publication has not been
profit, but rather the diffusion of useful informa-
tion, the present edition has been considerably
reduced in price, although, as will be seen in
looking over the index, I have made many additions which will probably be thought improvements.
But, as formerly, I warn my reader against
doctoring himself, except on an emergency.

Dec. 1849.   JOHN F. SOUTH.

PREFACE

TO THE FIRST EDITION.

Some time ago, for the purpose of aiding a literary institution which had been established in
"Our Village," I delivered some lectures, to one
of which I gave the name of Household Surgery,
its object being to afford useful hints as to
the means which people have in their own
power to employ when accidents happen which
require immediate attention and no medical man
is at hand, and often cannot be obtained for
hours. Such cases are neither few nor unim-
portant, and many serious consequences—nay,
even death—may be prevented if a judicious
person, having been put on the track, make use
of the simple remedies which almost every house
affords. Though the subject might seem an odd
one to select for a village auditory, yet there
was no difficulty in rendering it amusing as well
as instructive, and I believe I may justly say
that I did not fail of attaining my object. On
accidentally mentioning the circumstance to one
of my friends, who, by means of the printing-
press, furnishes literary food to a hungry and
never satisfied public, he said, "Why do not you

make a little book of it? Do; and I will print it." So we struck hands, and hence appears my small volume.

Now, no one who reads herein is to expect that he is to find a whole body of surgery, or that he will be fitted thereby to set up for an amateur surgeon capable of practising upon himself or his neighbours, for a vast deal must be passed by; not that there are any "secrets of the prison-house" to be kept, because I am one of those who think it would be well if people generally understood something about how they were treated and what they were treated for, as they would then be less open to the tricks of cunning knaves, whether enjoying the honour of a diploma or not, and more capable of estimating the value of honourable and upright practitioners, and therefore more ready to acknowledge obligations which can never be repaid, however many may be the fees given. But everything cannot be told, usefully at least, to general readers, for this simple reason—that were it set forth, they could not comprehend it. Still, however, there is much to be mentioned which may be of good service at a pinch, without interfering with the doctor; because I propose only showing how to manage, when and where he is not to be obtained, and the case is urgent. But if the doctor is to be had, let no one despise his privilege, but avail himself of it. Recollect what

the best book says: "Honour the *Physician**
with the honour due to him," Eccl. xxxviii. 1; and
again, "Give place to the *physician*, the LORD
created him," *ib*. 12. Whoever neglects this
advice, and doctors himself when he can be doc-
tored, is in much the same case as the man
who conducted his own cause, and had a fool for
his client.

As my work ran on, it struck me that I might
also render it more extensively useful by giving
some simple rules for the treatment of broken
limbs and other accidents, which are of not unfrequent occurrence on board ship and in the

---

* Under the term Physician must also be included
Surgeon, for in those days they formed but one faculty, as
we read, "Joseph commanded the *physicians*, his servants,
to embalm his father," which was certainly not physicians'
work. The apothecary indeed is specially mentioned, as
in "An ointment compounded after the art of the *apothecary*," Exod. xxx. 25; and also, "A perfume, a confection
after the art of the *apothecary*," ib. 35; but he does not
seem then to have had anything to do with the healing art,
although now, as the General Practitioner, he performs the
functions of all three branches into which it is now divided,
and that of the midwife to boot; and has the power, by Act
of Parliament, of compelling an aspirant to his particular
branch of the profession to undergo an examination and
obtain a licence to practise, which power physicians and
surgeons do not enjoy; and therefore her Majesty's lieges
may be killed and cut up by any pretender, provided they
do not send, but only prescribe, physic.

outskirts of colonies, where doctors are not to be had : and also at the same time that I might aid the Missionary in his high and holy calling, by enabling him to minister to the bodily ills of those among whom he is placed. The more common kinds of broken limbs require no great skill to discover, and if put into anything like their proper place and kept still, nature takes care of the rest, and the patient does very well. The rules I have laid down will be easily comprehended, and will answer in most cases.

It is very possible that some observations may be made on a Hospital Surgeon writing a book of this kind, intended for general use. I am very careless on this point, as I have had no unworthy object in view. The way in which, and the purport for which, the book has been written, are my apology, if any be needed, which I do not admit; and if I desired precedent, I need scarcely remind my readers how many of the ablest persons in science and art have held it no degradation to their high standing to render their particular branch of knowledge accessible, not only to adults but even to children, by cheerfully written works in simple language, often accompanied with homely illustration.

JOHN F. SOUTH.

CONTENTS.

xiv CONTENTS.

HOUSEHOLD SURGERY;

OR.

HINTS ON EMERGENCIES.

As it is impossible for any workman to practise his craft without tools, so it is needful that the person who, for the nonce, is to occupy the place of the Surgeon, should have

"all appliances and means to boot,"

wherewith he may be enabled to employ his small skill to the best advantage. On this account I think it advisable first to give some directions as to the making of poultices, washes, plasters, ointments, and other necessary things, which, taken together, and following the names-giving of the Father of Medicine and Surgery, HIPPOCRATES, I shall call

B

THE DOCTOR'S SHOP.

THE articles which all families, in the country, ought to be constantly provided with, are :—

Linseed meal ⎫
Poppy heads ⎬ For Poultices and Fomentations.
Camomile flowers ⎭

Sticking or Adhesive Plaster or Strapping (all names of one and the same thing) for bringing together and binding up wounds.

Blister or Spanish Fly Plaster.

Wax and Oil (soothing) ⎫
Spermaceti (same) ⎪
Turner's or Calamine (absorbent) ⎪
Oxide of Zinc (same) ⎬ Cerate or Ointment.
Yellow Basilicon or Resin (stimulating) ⎪
Citron or Nitrate of Mercury (same) ⎪
Lead (astringent) ⎪
Gall (same) ⎭
Sulphur

Opodeldoc or Soap ⎫
Camphor ⎬ Liniments
Mustard ⎭

Sugar or Acetate of Lead ⎫
Oxide of Zinc ⎪
Calomel ⎪
Sulphate of Zinc ⎬ For Washes.
Sulphate of Copper ⎪
Nitric Acid ⎪
Solution of Chlorinated Soda ⎪
———————————— Lime ⎭

Red Precipitate of Mercury.
Calamine Powder.
Powdered Galls.
———— Spanish Flies.
A stick of Lunar Caustic in a bottle.
Spirits of Wine
———— Turpentine   To be mixed with other
———— Hartshorn   things, and form washes,
Tincture of Opium, that is   liniments, and ointments.
    Laudanum

It is to be presumed that, for household purposes, there will never be lack of salt, mustard, vinegar, sweet oil, treacle, bran, oatmeal, flour, and bread, all of which, on one occasion or another, may be wanted for immediate use; and stale beer-grounds or yeast, which may sometimes be needed, are easily procured.

The materials for rollers or bandages, as linen or flannel, and broad tape ; for making pads, as flannel, blankets, bed-rugs ; for making pillows, as feathers, hay, straw cut into short pieces half-an-inch long, a substitute for chaff, if there be none handy, are to be found in most houses. In addition to these, a stock of lint (a quarter or half a pound) should always be kept.

Most families are also provided with the materials for making splints, as pasteboard, hat or bonnet boxes, and the flat thin boards upon which silk is commonly wound. But a common carpenter, or any one who can smooth a thin board with a plane, and can use a saw decently, will very easily make wooden splints of any size and form, or any other common contrivance which the case may require.

For making litters to carry persons who have met with severe accidents, hop-poles or other stout stakes, and blankets or horsecloths, serve; and a door or a hurdle is a litter ready made.

The Missionary or Emigrant should also be furnished with :—

A case containing six lancets.
A case of cupping instruments, containing—
Two cupping-glasses of different sizes ;
A spirit-torch, with cotton and a bottle of spirit.
A scarificator.
A pocket case of instruments, containing—
One common scalpel or surgeon's knife
——- Lisfranc's knife
——- Gum Lancet.
——- spatula, iron or silver.
——- pair of dressing forceps or tongs.
——————— small dissecting forceps for pulling out splinters.
——————— scissors, one end rounded.
——- silver probe.
——- caustic holder (silver).
——- tenaculum, or hook for taking up arteries.
Six curved surgeon's needles, in twos of different size.
A case of Tooth Instruments will also be exceedingly useful, containing—

Two pair of tooth forceps, one for children, the other for adults, and
—— key instruments fitted for the like purpose.

Thus far for the Surgical Department of the Doctor's Shop for home or foreign service.

But there are a few simple Family Medicines with which no house should ever be unprovided ; such are :—

A Bottle of Castor Oil.
Lenitive Electuary, or Confection of Senna.
Calomel.
Tartarized Antimony.
Iodide of Potash.
Sulphate of Quinine.
Ipecacuanha Wine.
Mercurial or Blue Pill.
Rufous or Aloes and Myrrh Pills.
Powdered Rhubarb.
————— Opium.
————— Gum.
Oil of Cloves.
—— Peppermint.
Carbonate of Lime.
Croton Oil.
A pair of grain Scales, with weights from half a grain to two drachms.
A glass drop-measure and a glass two-ounce measure.
A Dutch Tile, and a Spatula.

It is best to have all the fluids and powders in stoppled bottles.

Such, then, are the furniture of the Doctor's

Shop, both Surgical and Medical. Those who live in towns have no positive need for thus providing themselves, as doctors and physic are always in readiness, though even they may occasionally find the advantage of having medicine at hand. But to clergymen and others, whose lot has fallen among the pleasant places or country life, especially in those parts where medical aid is far distant, often unattainable for hours, and even if the doctor be luckily, in his long and toilsome round of duty, riding that way, hours must pass by before medicine can be obtained, the immediate administration of which might save with certainty (so far as human means avail) valuable lives, which are lost or put in extreme peril by the unavoidable delay of fetching medicine from a distance : to such country residents the Doctor's Shop I have recommended is invaluable. It is not expensive, it is not cumbrous. A shelf in a bed-room closet or a nook in the store-room will afford ample room for the whole. *But never forget that the shop is to be kept safely under lock and key, and that the key should never be given but to clear-headed and trustworthy persons who will not meddle with anything unbidden, as many of the materials used improperly are active and dangerous poisons.*

For those who like neatness, are fond of nicknacks, and have money to spare, Medicine

Chests of various sizes and price may be obtained from Apothecaries' Hall, or from most Druggists. But though in London it matters little whether the medicines themselves are obtained from the Hall or from respectable Druggists, as in either case they may generally be relied on, yet in country towns this is not the case; for there too often medicines remain in the druggists' bottles or drawers for so long a time, that they become good for nothing, although they have originally been of the best. Therefore *always have your medicines from Apothecaries' Hall, or from a respectable London druggist, desire to have the best, and do not mind the cost,—* no physic is better than bad physic.

Having thus stocked your Shop, we will next show how its contents are to be compounded into Poultices and so on.

FEW persons have not heard of JOHN ABER-
NETHY, one of the first Surgeons of his time, and
he to whom must justly be ascribed the praise
of having urged on those of his own particular
class in the profession (at a time when a Surgeon's duty was. held to be almost entirely confined to *cutting*), the importance of connecting
Medicine and Surgery together; for although
split, by custom, into these two branches, he
maintained that " Medicine is more one and in-
divisible than the French Republic." Upon
this foundation was his celebrity built; and his
example led Surgeons generally to study medicine as a powerful auxiliary to the practice of
their own particular branch, and no longer to
despise it and be thankful they knew nothing
about it, as a celebrated teacher once told his
class.

With a large and comprehensive mind, ABER-
NETHY did not despise the day of small things;
he did not think lightly of what many consider little matters in Surgery, not worth knowing to a practitioner, and still less fitting to

attract the notice of a professor and teacher. Hence his painstaking to impress on his pupils the importance of knowing

How to make a Poultice.

" Blessings or curses," as he used to say, " as they are well or ill made." And accordingly when Professor of Surgery at the Royal College of Surgeons, he described with great humour, as he was accustomed to his own private class, how this important branch of Surgical Cookery should be managed, and made the following apology for so doing, in his own peculiar but impressive style :—" A pretty fellow, truly, are you, to be appointed Professor of Surgery to the Royal College of Surgeons ; could you find no other subject better worth the attention of your audience than poultices? But I said, if any young Surgeon has had a troublesome disease to deal with, it would be good for him that he had been in trouble, for poultices are either blessings or curses, as they are well or ill made, and more commonly, as they are made, only irritate, instead of doing good."

According to this great Professor of the Art, poultices are of two kinds, the *evaporating poultice* or *local tepid bath*, and the *greasy poultice*, each of which serves a special purpose, and must be employed according to circumstances.

"The most soothing application for local disease," said ABERNETHY, "is tepid bathing, and this we can manage by putting on a poultice, being careful that the person is kept in bed, otherwise the poultice will serve the purpose of a cold bath. The poultice of which I am fond above all others, is

The Bread and Water Poultice, or Evaporating Poultice.

"And I tell you the mode of making it. Scald out a basin, for you can never make a good poultice unless you have perfectly boiling water; then having put in some hot water, throw in coarsely crumbled bread, and cover it with a plate. When the bread has soaked up as much water as it will imbibe, drain off the remaining water, and there will be left a light pulp. Spread it, a third of an inch thick, on folded linen, and apply it when of the temperature of a warm bath. It may be said that this poultice will be very inconvenient, if there be no lard in it, for it will soon get dry; but this is the very thing you want, and it can easily be moistened by dropping warm water on it, whilst a greasy poultice will be moist, but not wet."*

---

* ABERNETHY did not mean by this, that the bread and water poultice was wished to become dry, but rather that

A poultice .hus made, may be the *vehiculum crassum* of the doctors, the *stock* of the cooks, to be medicined or seasoned with laudanum, or poppy water, with carrot or horse-radish juice, or with decoctions of herbs, anciently known by the name of *stoups*, with which the patient or the doctor may be inclined to medicate it, instead of loading an already irritable and very sensitive part with a heap of hard poppy-shells, or scraped carrots and horse-radish, called poppy, carrot, and horse-radish poultices, but which increase rather than allay the sufferer's pains.

When vegetables—as carrots, horse-radish, and others—are used to medicate poultices, they should be bruised, put into a pot, covered with water, and simmered for about half an hour. The juice is then to be strained off and mixed with bread and water or linseed-meal, to the consistence of a poultice. The poppy fomentation may be used with bread or meal in the same way.

The Linseed-Meal or Greasy Poultice.

Is, on the authority already quoted, to be male in the following manner:—" Get some linseed powder, not the common stuff full of grit and

---

it should be *disposed to dry* by the evaporation of the water from it, and that as this occurred it should be again moistened so as to keep up the evaporating capability.

sand. Scald out a basin ; pour in some perfectly
boiling water ; throw in the powder, stir it round
with a stick, till well incorporated ; add a little
more water and a little more meal : stir again,
and when it is about two-thirds of the consistence
you wish it to be, beat it up with the blade of a
knife till all the lumps are removed. If properly
made, it is so well worked together, that you
might throw it up to the ceiling, and it would
come down again without falling to pieces ; it is,
in fact, like a pancake. Then take it out, lay it
on a piece of soft linen, spread it the fourth of
an inch thick, and as wide as will cover the
whole inflamed part ; put a bit of hog's lard in
the centre of it, and when it begins to melt, draw
the edge of the knife lightly over and grease the
surface of the poultice. When made in this
way, oh ! it is beautifully smooth : it is delight-
fully soft ; it is warm and comfortable to the
feelings of the patient. So much for the greasy
poultice."

The Bran Poultice

Is a sort of " entire," or half-and-half, partly
poultice, partly fomentation ; and is a very good
application for setting up and keeping up perspiration on a part ; but it requires to be often
changed, for it very quickly becomes sour, and

then has not the most agreeable smell. It merely consists of bran moistened, but not made wet, with hot water; and enough of it should be put into a flannel bag, sufficiently large to cover the part, to fill it about one-third; if more bran be put in, the bag becomes unpleasantly heavy. It must then be held before the fire, and the bran turned about again and again till it is thoroughly heated. Thus warmed, it must be quickly applied, and the bran should be gently spread, so as to cover the whole extent of the bag.

Stimulating Poultices

Are required for two purposes—either to hasten the separation of a dead part or slough, or as it is called in vulgar language, "a setfast," or " core;" or to irritate the skin where it is inconvenient to apply a blister, or for the purpose of rendering the operation of a blister more speedy. For the first of these objects, yeast, stale beer-grounds, or treacle, is used; for the second, mustard.

Yeast Poultice

Is made by mixing a pound of flour, or linseed-meal, or oatmeal, with half a pint of yeast or beer-grounds. The mixture is to be heated in a pot, carefully stirred, to prevent burning, and when sufficiently warm, must be spread on linen like any other poultice.

Treacle Poultice

May be made according to the same proportions, heated and applied in the same way.

Mustard Poultice

Is a most excellent and safe application. It is generally recommended to make it by mixing half a pound of mustard with the same quantity of linseed-meal or oatmeal, and sufficient boiling vinegar, to a poultice consistence. Some use only mustard and boiling water, making the poultice of the usual stiffness.

Now it so happens, I have had, for many weeks together in my own person, nightly experience of mustard poultices, and have found mustard and water quite sufficient for the purpose required; and cold water will do as well as that which is warm, though the latter, however, is more agreeable to the feelings at first than a cold mustard poultice; but the discomfort from the cold ceases in the course of two or three minutes, and soon is there heat enough, and to spare. If you wish to have a mustard poultice act quickly, mix it with hot or cold water as you please, as thick on.y as you would have it mixed for the dinner table, and you will have no reason to complain of it not performing its duty well.

Two things are to be remembered in applying a mustard poultice. Do not let it be applied immediately to the skin, for after it has remained on the proper time, a quarter or half an hour, which will inflame the skin sufficiently, the part will be so exceedingly tender, that the removal of the poultice, with the handle of a spoon and sponge, will be a very difficult matter, and unless it be got off entirely, the patient will suffer a martyrdom. Therefore, spread the mustard about a quarter of an inch thick on a piece of fine muslin, and put the muslin next to the skin, and the watery part will act well through the muslin. The second thing to be done, is, after the poultice has been removed, to clear off quickly and lightly with a soft warm wet sponge whatever of the mustard remains on the skin, which must then be gently dried with a soft handkerchief.

If a mustard poultice be put on a child, it should be taken off two or three minutes after the skin reddens.

Cold Poultices,

Made with bread and cold water, or bread and lead-wash or Goulard's water, as it is com-

monly called, are favourites with some people,
but not with me. For they are not only disagreeable from their coldness, at first—which
however soon ceases after being worn a short
time, as they become tepid ; but also from their
harshness. If ever used at all, they should be
made with hot water, which renders them soft,
and when they are cooled may be applied ; but
I like them not.

OF FOMENTATIONS.

FOMENTATIONS are warm fluids applied for the purpose of encouraging perspiration on the skin, and thereby to diminish inflammation, and to render the skin yielding, so that the swelling which accompanies inflammation may be less painful, by the greater readiness with which the skin yields, than when it is harsh and dry. For the same reason, that of encouraging the skin to yield, fomentations are good applications to bruised parts, in which more or less blood escaping from the vessels that have been burst by the bruise, puffs up the skin and produces swelling of the injured part.

The common mode of using a fomentation is to dip into it a sponge or piece of flannel, and then either to squeeze the fluid from it over the part, or to pass the sponge again and again over it, making slight pressure till the sponge be emptied; and then dipping it into the fomentation, to repeat the same proceeding again and again, for the space of a quarter or half an hour; which done, the part is said to be well fomented, or " fermented," as the nurses call it. And truly a pretty ferment it is; for the patient's imme-

diate neighbourhood is deluged in wet; and if
from circumstances the operation have been performed in bed, the bed-linen and blankets, and
the bed itself, are sopped through and through,
and the patient obliged to lie in a wet, uncomfortable state, from which he runs a fair risk of
catching severe cold, or he must be moved into
another bed, which is not always convenient.
And even if no mess be made, which is scarcely
possible, the person already suffering from pain
is wearied with the continued exposure and frequent dabbing with the sponge or flannel, and,
if fortunately he escape harm, does not derive
half the benefit he would by properly using the
fomentation.

The *proper mode of fomenting* makes comparatively little mess, and does not in the least
fatigue the patient. A common jack-towel, three
or four fold thick, or what is still better, if at
hand, a piece of oiled silk or cloth, or even a
leather, should be smoothly spread on the bed,
beneath the part to be fomented. Three or
four pieces of thick house-flannel or of blanket
are to be cut, each of sufficient size to enwrap
completely the limb, or to cover the belly or
chest, whichever be the part to be fomented.
Another piece of oiled cloth, or two or three
thicknesses of jack-towel, of rather larger size
than the flannel, are also to be provided.

The fluid to be used as fomentation should be put in a pan or pail, and heated as hot as feels comfortable to the part on which it is to be applied. It is usual to recommend it of such warmth as the hand will pleasantly bear ; but this is a very careless proceeding, for nine times out of ten the hand of the person who applies the fomentation has been accustomed to hot water a few degrees below scalding, and bears it with as little inconvenience as the blacksmith picks up a hot horse-shoe which would burn to the bone the hand of one not accustomed to handle hot iron. The consequence, then, of this hand-trial of the heat of the liquid is, that though the fomenter finds it pleasantly warm, the fomentee feels it scalding hot. This seemingly trifling point must, therefore, not be forgotten ; nor must it be overlooked that the heat of the fomentation should be kept up by continually adding a little more of the hot fluid, as may be needful.

All things being now ready, the business begins by plunging the flannel or blanket into the fomentation, in which it should remain covered over till thoroughly soaked, and heated to the heat of the fluid. At the first dipping, the fomentation must be hotter than afterwards, because the flannels, which are cold, will lower it, and therefore not get the desired warmth. In about five minutes the flannel will have soaked and warmed thoroughly ; one piece must then be

taken out, wrung dry as quickly as possible by
two persons, spread out, and wrapped round or
laid upon the part to be fomented, and then the
flannel quickly overwrapped with oiled silk or a
jack-towel, and the limb gently laid down on the
bed. A flannel thus managed will keep its
warmth for ten minutes or a quarter of an hour,
or even longer, and may then be immediately
replaced by another soaked flannel, which has
been wrung out and made ready while the first
is being removed. In this way a part may be
fomented for hours without the least fatigue to
the patient, or chance of his catching cold; and
there is not a part of the body which cannot in
this way be easily and agreeably fomented.

Occasionally it happens that very large sur-
faces of the limbs or body may be advantaged
from the application of moist heat, and a poultice
for that purpose is not unfrequently ordered; but
when a poultice is large, if of linseed meal, it is
very heavy, and if of bread, both heavy and
wet; both are liable to smear about the bed-
clothes, and as the crumbs get dry they render
the patient very uncomfortable by sticking into
him. In such cases a fomentation, applied as
directed, in every respect answers the same purpose, is more manageable, and without any of
the inconveniences of the poultice.

Some parts of the foot and leg, and the hand
and fore arm, can be most easily and effectually

fomented by putting the former into a pail, and the latter into a tongue-pan or foot-pan, either of which is to be filled with the warm fomentation sufficiently to cover the part to be fomented. The heat of the fluid must be kept up by occasionally pouring in more which is hot.

DIFFERENT KINDS OF FOMENTATIONS.

THE fluids used for fomenting are of various kinds; and most people have some one or other to which they are specially attached.

Warm Water

Is the most simple, most ready, and oftentimes as useful a fomentation as can be employed.

Poppy-Water.

Take four ounces of dried poppy heads; break them to pieces, empty out the seeds, and put the shells into four pints of water, boil for a quarter of an hour, then strain through a cloth or sieve, and keep the water for use.

Mallow-Water

Take four ounces of dried mallows, and boil in four pints of water for a quarter of an hour, and strain.

Sometimes two ounces of camomile flowers are boiled with either of these; but I do not know that they are of much service.

Vapour Bathing

Is often of great service in stiffness and swelling of joints or of other parts, and may be managed with little difficulty, with apparatus which may be constructed of common household utensils. The part to be steamed must be placed under a cradle, made as described at p. 215; or under a light wicker basket of sufficient size to cover without being close to the part. Over either of these a thick blanket or two is to be thrown and carefully pressed around it, so as to make a close chamber. To furnish the vapour, a funnel, either earthenware or tin, must be turned with its wide part over the top of a large saucepan or of a tea-kettle, from which the lid has been removed, and both bound together with a thick soft cloth. The tube of the funnel must be inserted into a piece of lead pipe about eighteen inches long, which must be curved and the two bound together with soft cloth; and to the other end of the lead pipe, a hollow cane, or joint of a fishing-rod, a yard in length, must be attached. By this contrivance the saucepan or kettle being kept boiling on the fire, the steam can be readily conveyed under the cradle or basket, the end of the hollow cane having been passed beneath the covering blanket at the greatest distance from the part to be steamed. Care must be taken

not to get up the steam too fast, nor to keep the
water boiling too briskly, nor to direct the end
of the tube so that the steam should rush against
the ailing part, as the result of inattention to
these circumstances will be scalding. The water
in the pan or kettle should be made to boil and
the steam allowed to stream forth before the end
of the pipe, which should have been previously
so arranged that it can be directed towards the
top of the basket, is put beneath the blanket.
This having been done, it will be sufficient to
keep the kettle gently boiling or simmering to
furnish the necessary quantity of steam, and five
or ten minutes will be long enough to continue
the process.

If it be inconvenient to bring the patient near
the fire, the funnel may be turned over the top
of a tea-urn (without its lid) which has been
filled with boiling water and kept boiling with its
red-hot heater.

A DRY FOMENTATION.

THIS may seem a very paradoxical expression;
but "what's in a name," if it convey what is
meant? Dry heat is occasionally very useful,
and the dry fomentation furnishes it easily, and
retains it.

A thin flannel bag must be made rather larger
than the part to be covered, and this is to be half

filled with camomile flowers or hops, and then sewn up. Thus prepared, the bag may be held before the fire, turned from side to side, shaken up again and again, till the contents be thoroughly heated. Or it may be equally well managed, and sometimes more conveniently, by putting coals in a warming-pan, and passing it over the camomile bag again and again till it be sufficiently heated. The bag must then be quickly applied to the part, and covered with a napkin. It is always well to have a couple of camomile bags in use at the same time, so that one may replace the other as soon as that which had been put on begins to feel cool.

Another dry fomentation is made by filling a bag with salt, and holding it before the fire till thoroughly heated. It is then applied as warm as the person can bear, but it is objectionable on account of its weight.

LOTIONS OR WASHES

ARE either cooling and soothing, or irritating and stimulant, or drying and absorbing.

COOLING OR EVAPORATING WASHES.

THE object of these washes is to lessen the inflammatory condition of a part by diminishing its increased heat, which is one of the signs of inflammation. In using an evaporating wash, care must be taken that the linen which has been wetted with it shall be freely exposed to the air, so that the fluid may evaporate as it is heated by the inflamed part, or by the mere heat of the body. It is therefore clear that a part thus treated must neither be wrapped up in three or four folds of linen, nor carefully covered up with the bed-clothes. A single piece of linen should be well wetted in the wash and laid on the part ; and as it dries by evaporation, must be again wetted, either by taking off, dipping afresh in the wash, and putting on again, or by filling a sponge with the wash and squeezing it over the linen without disturbing it, which is the better mode. The part thus covered with the wet linen must have no further covering, and the bed-clothes should therefore be turned off.

Cold Water.

Is as good a wash as any to produce evaporation, if care be taken to have the wet linen well exposed to the air ; and it has the further advantage of being almost always at hand.

Spirit Wash.

Half a quarter of a pint of spirits of wine, or a quarter of a pint of brandy, or any other good spirit, added to a pint of water, make this wash.

Vinegar Wash.

Is made by mixing one-fourth of vinegar to three-fourths of water.

To a pint of either of the former washes half a tablespoonful of laudanum may be added, if the pain suffered be very severe.

Lead Wash, or Goulard-Water or White Wash,

As it is often called, for common purposes may be made by dissolving one drachm of sugar of lead in a pint of soft water. Some persons are very fond of using this wash, with the addition of spirits of wine, as an evaporant ; but I do not like it, for it renders the skin very dry and harsh, and its sedative virtue acting through unbroken skin, I do not hold of much value. Under other circumstances it is often very useful.

When used as a wash for the eyes, two grains of the sugar of lead are to be dissolved in two tablespoonfuls of water.

STIMULATING WASHES

ARE employed for encouraging sluggish sores to heal. They are usually applied by dipping lint into them, which, being then put on the sore, is confined with a roller.

Black Wash

Is a most valuable application : it is composed of a drachm of calomel in half a pint of lime-water,* and should always be well shaken before using, as much of the calomel drops to the bottom of the bottle.

This wash is often rendered more serviceable by adding a teaspoonful of laudanum to the above quantity; and sometimes in very irritable sores one ounce or two tablespoonfuls of mucilage or moderately thick gum-water may also be added with advantage.

Nitric Acid Wash

Is very proper for a sluggish sore with a stinking discharge, which often depends on the bone being affected. It may be used in the proportion of one drop of the acid to a tablespoonful of soft water.

---

* How to make lime-water will be shown at p. 28.

Chlorinated Soda Wash

Is also a very capital wash for purifying stinking ulcers or sloughy wounds, which it also gently stimulates. It may be used in the proportion of two tablespoonfuls of the solution of chlorinate of soda, commonly sold as the Disinfecting Fluid, to half a pint of water.

Chlorinated Lime Wash

Is made according to the same proportions.

DRYING WASHES

ARE used for cracked skin on the legs, or to promote the scabbing of large ulcers, more especially after scalds and burns which discharge very freely.

Lime-Water

Is a very simple application, is one of the best, and very easily made. Take half a pound of unslaked lime, and three-quarters of a pint of water. Put the lime into an earthen pot, and pour a little of the water upon it, and as the lime slakes pour the water on by little and little, and stir up with a stick. The water must be added very slowly, otherwise the lime will fly about in all directions, and the great heat suddenly pro- duced will crack or break the vessel which contains it. After three or four hours, when the slaked lime has sunk to the bottom, the clear fluid may be poured off, and put in a stoppled bottle away from the light.

Oxide of Zinc Wash

Is made by putting four drachms of oxide of zinc into a pint of lime-water, which does not, however, dissolve, but merely suspends it. It is, therefore, always necessary to shake the bottle well up, so that the linen may entangle the proper quantity of the oxide.

LINIMENTS

ARE used for promoting the removal of swellings in various parts, by encouraging the action of peculiar vessels called absorbents, the duty of which is to remove from the body all those portions of it which have become useless or even hurtful, so that new materials may be supplied in their place by the blood-vessels. Liniments are also used to excite irritation on the skin, and produce a diversion to it of inflammation going on in deeper-seated parts.

All liniments are applied by rubbing, either with the bare hand, or with the hand wrapped in a bladder or piece of oiled silk, if the liniment be very sharp and likely to affect the hand of the rubber, as some kinds will do.

Simple Rubbing with the hand or *with the hand covered with a hair glove* is generally sufficient to quicken the removal of swellings of the limbs, which often remain after broken bones have been released from the bandages, or which may have arisen from other causes. It is not needful that this operation should be performed by rubbing "as hard as they can," as it is commonly called, which is very wearying to the rubber and to the rubbed. All that is necessary, if the swelling be of the whole limb, is to run the flat of the hand, with gentle pressure, up and down for ten minutes or a quarter of an hour at a time. But if a joint have to be rubbed, the method practised many years since by GROSVENOR, of Oxford, who employed all the old women in the town for the purpose, is the best; it consists in placing the two hands, one on each side of the joint, moving them alternately up and down at the same time, and making each hand travel half round the joint.

Sometimes *Neat's-foot Oil* is rubbed in upon the joint by either of these modes, and has the advantage of rendering the movement of the hands more easy and allowing the rubbing to be continued a greater length of time.

Hartshorn and Oil,

In the proportion of one-third of the hartshorn to two-thirds of oil, is a very good stimulating liniment for stiff neck and lumbago.

Camphor Liniment

Is made by rubbing down an ounce of camphor in four ounces of olive oil.

Soap Liniment, or Opodeldoc,

Is a great favourite. If not bought, it may be easily made by slicing three ounces of hard white

soap and an ounce of camphor, putting them into
a bottle and adding half a pint of spirits of wine
or brandy, or any other strong pure spirit, and as
much water. The bottle should be shaken from
day to day, till the soap and camphor are dissolved, when the liniment is fit for use. If it he
desired to lull violent rheumatic pain, a teaspoonful of laudanum added to two tablespoonfuls of this liniment will be very beneficial.

Mustard Liniment

Is, for stimulating the surface, the very best of
all, as it is very manageable and may be made
to act either very slightly or so severely as to
take the skin off, according to the quantity
used and the time the rubbing is kept up.
The best guide as to the quantity required
is the feelings of the person rubbed. At first
there is a pleasant sensation of heat, then a
little pricking, and next a positive smarting;
when this is produced, leave off, for if you
continue to rub you will soon flay the skin, and,
as a consequence, not be able to rub again for
three or four days. An ounce of fresh flour of
mustard put into a bottle with a pint of spirits
of turpentine, and shaken daily for two or three
days, make this liniment. The mustard will
settle to the bottom, and the .clear fluid should
be then poured off. Do not leave the mustard

in and shake the bottle up before using; if you do, you will give the skin a coating of mustard and render the application unnecessarily severe. It is the best thing for lumbago and chilblains.

Sometimes it is necessary to keep up irritation on the skin for a length of time without disturbing the constitution, which some irritants will do. The best application for this purpose is—

Croton Oil,

Of which ten or a dozen drops are to be rubbed in lightly with the fingers, guarded with a piece of oiled silk, for two or three nights. Generally on the second day the surface is red and puffy, and on the third day a large crop of little blisters about the size of hemp-seed cover the skin. When these appear the rubbing must be stopped. In the course of a few hours the fluid in the blisters changes to matter, and these pustules begin to tingle and itch furiously. As soon as this happens, prick each with the point of a needle and press out the matter with a handkerchief; or if you have not patience for this, run the back of your nails smartly over them in various directions, and crack the pustules as fast as you can. The sensation is rather agreeable than otherwise; and soon after, the luxury of a good scratch is

D

beyond all price. Sometimes a few of the
pustules are so tough that they cannot be thus
broken, and, unless pricked and emptied, become
excessively painful, forming a sort of small
boil, which the sooner it is got rid of the better.
In the course of a week the skin has been completely reproduced, and then the croton oil may
be used again; but it does not blister quite so
quickly as when first applied. The croton may
be used for months, and is a most excellent
mild irritant, and infinitely superior to that most
cruel and inhuman of all human inventions, a
perpetual blister, as it is called; that is, a blister
from which the skin having been cut off, the raw
surface is dressed with savine ointment. I am
sure no person who has ever smarted under this
awful torment would order it for another.

OINTMENTS.

THE base of all ointments is grease, and they are used for dressing wounds and sores, to prevent the sticking of the lint or linen with which they are covered, and to protect them from the air and from filth: the most simple kinds serve this purpose best. But sometimes medicine of various kinds is mixed up with grease to form ointment, through the means of which the medicine acts on the surface of the sore. It is necessary there should be different modes of dressing sores with medicine, as they are very capricious: one sore will bear an ointment, but neither wash nor poultice; another will be quiet only with a wash; and sometimes the same sore will do well with a medicine, at one time in an ointment, and at another in a wash.

Simple Ointment

Is made by melting in a pipkin by the side of the fire, without boiling, one part of yellow or white wax, and two parts of hog's lard or olive oil.

Spermaceti Ointment

Consists of a quarter of an ounce of white wax, three-quarters of an ounce of spermaceti, and

three ounces of olive oil melted as before.
This is the common dressing for a blister.

Elder-flower Ointment

Is the mildest, blandest, and most cooling ointment, as the old women call it, which can be
used, and is very suitable for anointing the face
or neck when sun-burnt. It is made of fresh
elder-flowers stripped from the stalks, two pounds
of which are simmered in an equal quantity of
hog's-lard till they become crisp, after which the
ointment whilst fluid is strained through a coarse
sieve.

For stimulating wounds or sores either of the
following ointments is used :—

Red Precipitate Ointment

Is made by mixing thoroughly a drachm of finely-
powdered red precipitate of mercury with an
ounce of simple ointment.

Resin Ointment, or Yellow Basilicon,

Is composed of two ounces of yellow wax, five
ounces of white resin, and seven ounces of hog's-
lard ; these must be slowly melted together, and
stirred constantly with a stick, till completely
mixed. This ointment is sometimes used in
treating scalds and burns; also for dressing

blisters, when it is wished to keep up a discharge from them for a few days.

As there are drying washes, so are there drying ointments, of which the two following are the best :—

Calamine Ointment, or Turner's Cerate,

Consists of one half a pound of yellow wax and a pint of olive oil, which are to be melted together ; this being done, half a pound of calamine powdei is to be sifted in and stirred till the whole be completely mixed.

Zinc Ointment

Is made by rubbing well together one ounce of oxide of zinc and six ounces of hog's lard. It is commonly used for dressing the sores remaining after scalds and burns, to lap up the great discharge which generally follows ; and it is a very good application to cracked skin, from which a watery fluid oozes and irritates the neighbouring skin.

Iodide of Potash Ointment

Is a very excellent and safe application for enlarged glands in the neck, or elsewhere, so long as the skin over them be not red or inflamed. It is better smeared on thickly at night and covered with a piece of linen, than rubbed ; for if

rubbed it sometimes irritates the skin, and must
be left off.   It is made by dissolving a drachm of
the iodide of potash in a little water, and then
rubbing it well together with a knife with an
ounce of hog's-lard.

Lead Ointment

Is made by mixing one drachm of sugar of lead,
which must first be rubbed into fine powder with
a spatula, with one ounce of lard.

Gall Ointment

Consists of a drachm of powdered galls, fifteen
grains of powdered opium, and an ounce of lard,
mixed well together with a spatula.

These last two ointments are commonly used
for piles.

The list of ointments cannot be concluded
without the following, a certain cure, if properly
applied, for that nastiest of all nasty, and most
easily-caught disease, the Itch, which, although
generally found among poor people, occasionally
steals into the house of the wealthy, and, like a
snake in the grass, is often only accidentally
discovered :—

Sulphur Ointment

Is made by rubbing well together three ounces
of flowers of sulphur and half a pound of hog's-

lard. The proper mode of managing it is, for the miserable infected to rub himself well *all over* with the ointment night and morning for three days, during which time he must wear, without change, some old body linen, stockings, and gloves, and lie in a pair of old sheets or blankets. Washing in the least degree is to be as carefully avoided as the plague, for it will protract the cure. On the fourth day let him go into a warm bath, wash himself clean, and most generally he will then be found quite well. But if not, the sulphur-soaking must be continued three days longer. Everything which had been worn during the cure should be burnt, sheets and all; but the blankets may be scoured.

Our northern neighbours have a

Compound Sulphur Ointment,

Which is also an excellent remedy for this abominable nuisance, and by some thought better than the simple sulphur ointment. It is composed of half a pound of flowers of sulphur, two ounces of bruised hellebore-root, one drachm of nitre, half a pound of soft-soap, and a pound and a half of hog's-lard, scented with thirty drops of essence of bergamot, and all well mixed together.

PLASTERS

ARE used for keeping wounds together and for binding up sores. People not accustomed to it cannot make plasters, but they may have them from the druggist in the roll, and easily manage to spread them on linen with a hot knife or spatula, which should be sufficiently heated to slightly brown white paper, but not hotter. Plasters are best spread when wanted, and used whilst fresh; for if unused till after they have been some weeks spread, the plaster flakes off the linen, unless more resin be added to the plaster at first than is proper. The best material for spreading plaster on is moderately thick glazed calico, which is less expensive and cumbersome than white leather. The piece of calico to be spread having been cut off, and strained upon a board by fastening at each end with tacks, some of the plaster, which has been melted slowly in a little iron pot or pipkin by the side of the fire, but not on it, should be poured across one end of the calico, and then with the edge of the warm knife run lightly and quickly to the other end, supplying a little more warm plaster whenever that already put on has been spread. In this way, with very little practice, plaster can soon be smoothly and nicely spread.

Some tough plasters, as Burgundy pitch, may be spread on white-brown paper; such is the way in which the so-called Poor Man's Plaster is made.

The Common Sticking-plaster or *Strapping*

Is made by melting together two parts of soap-plaster and one of resin-plaster; the latter being added to make it sticky : but if fresh spread, the soap-plaster alone will stick well enough. It may be spread either on calico or on black silk, to render it more sightly.

Court Plaster, or *Black Sticking-plaster*,

Is made by brushing pretty thick gum-water over black silk strained tightly. After having been dried, it will keep a long while if not exposed to damp. As is well known, it merely requires moistening with the tongue to fit it for use, and answers very well for slight cuts.

Isinglass Plaster

Has of late come much into vogue, and will answer the purpose of sticking-plaster if that cannot be obtained, to which, indeed, some think it even preferable. This is made by dissolving isinglass in a small quantity of boiling water, and then mixing it with sufficient spirits of wine to keep it fluid whilst gently brushed over silk or fine linen.　As the spirit evaporates, the isinglass forms a glaze on the silk or linen.　It also must be kept dry, and when required for use, strips of needful size may be cut off, and immediately before applying must be quickly and lightly brushed over with a hot moist sponge, which dissolves the glaze sufficiently to make it sticky, in which state it is put on, and almost immediately sticks fast to the part on which it is placed.

Blister, or *Spanish-Fly Plaster*,

Is rather an ointment than a plaster; and is sufficiently soft to be spread with the thumb, as the heat of a knife spoils its blistering property.　It may be made by dissolving in a pot by the fireside three ounces and a half of white or brown wax, the same quantity of suet, an ounce of resin, and three ounces of lard.　These are to be heated, only sufficiently to dissolve and allow of being well mixed by stirring together with a stick. Six ounces of Spanish flies finely powdered are then to be sifted in, and the stirring continued till the whole be completely mixed.　The plaster thus made should be poured into another pot to cool, when it is ready for use.

BLOOD-LETTING.

BLOOD-LETTING is required under many conditions of accident and disease. It is generally
effected by leeches, cupping, or the lancet; and
the use of the latter is that most commonly
known as bleeding or blood-letting, or breathing
a vein, as it was called in the olden time. One
of these modes is sometimes preferable to the
other, according to the particular circumstances
of the case ; but on the whole, cupping has the
advantage over the other modes, inasmuch as by
it not only can a *certain* quantity of blood be obtained very quickly, and without exhausting the
patient, as is too commonly the case with leeches,
but also the blood can be at once got from
the particular part whence it is desired to be
drawn, generally without knocking down the patient's strength, as frequently happens in bleeding
from a vein, from which, indeed, the necessary
quantity sometimes cannot be obtained in consequence of the person fainting.

LEECHING.

COMMON as leeches are now,* few persons have any notion of the distance from which they are brought for our use. Our own country furnishes at present few, if any, medicinal leeches. Formerly they were imported from France, but now many are fetched from Syria, and, as they are very delicate creatures, vast numbers of them are often lost in a rough passage across the sea.

The *Medicinal leech* (*fig.* 1) has three horny teeth in its mouth (*fig.* 2, *a a a*); it is marked with six orange-coloured stripes running from the head along the back and sides to the tail; its belly is steel-blue,

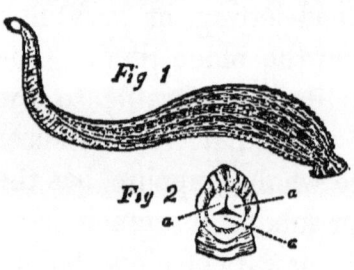

with regular yellow spots, which are often so numerous that they seem the ground-colour, and the steel-blue the spots; in rare instances the yellow spots are wanting, and the whole belly is steel-blue. These marks distinguish it from the *horse-leech*, of which the back and sides are blackish brown or blackish grey, without any marking, and its belly yellowish grey; and from

* I can well recollect when they were sold for two shillings or half-a-crown a-piece, thirty years ago.

the *common leech,* which is light brown, spotted with black above, and its belly greyish brown.

Leeches should be kept in a cool place, in a stone or glass jar filled with rain or river water, and tied over with coarse linen to prevent their escape, though it allow them air. The water should be changed only when it begins to foul, as too frequent disturbing destroys them. They are also sometimes found dead after storms.

There is often a good deal of trouble in getting leeches to fix. The part on which they are to be applied should be carefully cleared of perspiration, and wiped with a cool moist cloth so as to leave it damp. If they do not take readily, the part may be moistened with a little sugar and water, or sweet beer. But if this do not answer, the skin may be gently scratched with a needle-point till the blood come, and then they will generally take.

If it be wished to apply the leeches as near as possible on one spot, the best plan is to put them all in the deep part of a small pill-box, or in a small wine-glass, which is to be turned down on the part. If required to be spread over a large surface, as upon one of the limbs or the belly, they must be put on singly and by hand, which is often a very tedious and tiresome job. They should then be held tightly by the tail, wrapped in a piece of wet rag, so that they may

be less inconvenienced by the heat of the hand ;
and if the leech do not soon fix, it is best to put
it into the water to cool itself, and after applying
some others to try it again.   It is always best to
have more leeches than the number directed, in
case some will not bite.

When the proper number have been applied,
they should be left quite alone, or they are apt
to unfix, and, wandering about, are of no further
use. When they have sucked their fill they
generally drop off, and should then be put in a
plate and sprinkled with a little salt, which quickly
makes them throw up the blood ; and so soon as
they have emptied themselves they should be put
into plenty of fresh cold water, so that they may
get free of the salt, for if left in it, or if too much
be put on them, they contract violently, and die
almost immediately.

After all the leeches have come off, the bleeding from the wounds is to be encouraged, by
first sponging off smartly whatever clotted blood
there may be, and then covering the part with a
warm bread-and-water poultice, which must be
changed every half-hour, so long as it may be
thought necessary to keep up the bleeding.
This is much better than leaving the surface exposed and mopping with a warm sponge, which
is very fatiguing to the patient, besides exposing
him to the danger of catching cold.

One disadvantage in the use of leeches is the great uncertainty as to whether too little or too much blood is obtained by them. Sometimes, after much trouble in getting them to take, they will either only half fill themselves, or after they have come off the wounds will scarcely bleed at all. One wise man, to remedy this inconvenience, has proposed cutting off the tail, so that by this artful dodge, the blood continually flowing from the leech, it may be inveigled into the notion that its stomach is unfilled, and therefore would continue sucking for ever. This is, of course, downright nonsense; and one might be surprised at any such absurd proposal, were it not certain there is no lack of simple people to follow this or any other foolish advice.

Getting too little blood, however, is a matter of very trifling consequence in comparison with getting too much, for instances have occurred in which leech-bites have continued bleeding for days in grown-up persons as well as in children, bringing them into a very dangerous condition; nay, there is no want of well-authenticated cases of death caused by bleeding leech-bites, and that too in the course of twenty-four hours. The cause of this serious business sometimes is a peculiar kind of constitution, in which the blood will not clot sufficiently firm to stop the bleeding; or it may be some little artery has been wounded by the bite in such way that it cannot be stopped by a clot of blood.

If, then, a leech-bite continue bleeding for some hours, and the person, more especially if an infant, begin to be very faint, and the countenance and lips pallid and cold, like marble, no time must be lost in stopping the bleeding.

This is best done by thrusting a moderate-sized darning-needle (*a*) into the skin on one side of the bite (*b*), and bringing its point well out on the other side.   The whole

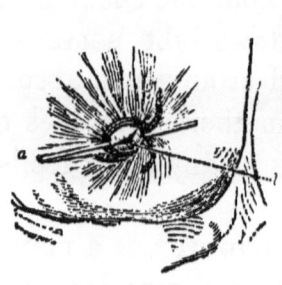

wound is thus lifted up, and a piece of strong silk or thread is then to be tied or wound round it beneath the two ends of the needle, and this in general effectually stops the bleeding.   This is much better than thrusting a fine-pointed piece of lunar caustic into the wound, as is often, but not always successfully done.   After three or four days the thread may be cut and the needle carefully withdrawn, and there usually the matter ends.

But in the peculiar state of constitution which has been noticed, sometimes even after the removal of the needle and thread the bleeding will continue.   Nothing then. remains but to thrust into the bottom of the wound a bit of thin iron wire heated white-hot, which I have never known fail to stop the bleeding. Though this may seem a very horrible proceeding, it is not very painful if the iron be white-hot, as it destroys the sensation in an instant; but whether it give pain or not is a matter of no consequence, as it is the only sure mode of saving the patient.

CUPPING

Is the most certain way of obtaining precisely the quantity of blood required: it has the advantages of being easily learnt, and of very rarely being followed by ill consequences. It is, therefore, the safest method to be followed by an unprofessional person.

The regular apparatus for cupping consists of cupping-glasses (*fig.* 1), of two or three sizes; a scarificator, or box of lancets (*fig.* 2) which are made to shoot out and return with a spring; and a spirit-torch (*fig.* 3), which is a small hollow metal globe filled with spirits of wine, with a long tube stretching from it and filled with some cotton threads which pass into the globe, and from it are fed with spirit.

Fig. 1.   Fig. 3.   Fig. 2.

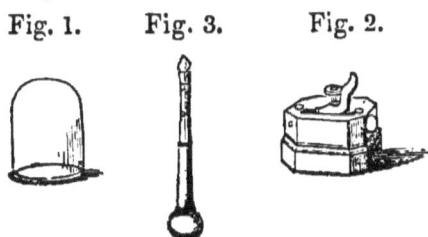

The place of cupping having been determined on, and the glasses, one or more to be applied, being chosen, each glass is to be placed with its mouth downwards, upon the part, on one side of its edge, leaving, however, sufficient room for the 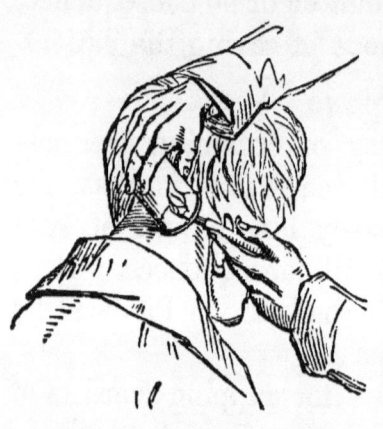 entrance of the pipe of the spirit-torch; which, being lighted, its end is to be quickly pushed in, taking care, however, to hold it so that it shall not burn the skin, which is the most troublesome and difficult part of the operation. The pipe must be held a few seconds in the glass to rarefy the air; then quickly withdrawn and the cupping-glass as quickly brought down upon the skin, so as to apply the whole of its edge. Immediately the torch has been withdrawn, and the glass completely 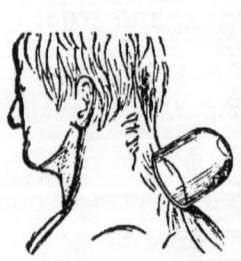 turned down, the rarefied air within recovers its ordinary condition; and a vacuum or space unoccupied by air, being thus formed, the skin draws into the glass, and fills the vacancy. The glass does not fall off, as the weight of the external air holds it firmly down.

In this state it must be left two or three minutes, by which time the indrawn skin will become very deep red, or almost blue, in consequence of being gorged with blood, which is prevented escaping by the pressure of the edge of the glass. Now is the time to remove the glass, which is done by grasping its upper part with one hand and inclining it a little on one side, whilst the thumb of the other hand pressing the skin firmly near its edge, the air rushes in with a hiss, and the glass comes off. Immediately this is done, the scarificator—its lancets having been set—is to be placed with its face upon the skin which has been drawn up;

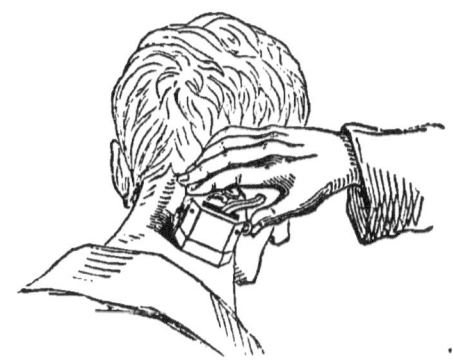

and then, by touching its spring, the lancets make their cuts and return home. The cupping-glass is now again put on, in the same way as at first and upon the same part, and very soon the blood is seen oozing up through the wounds, and fills the cupping-glass more or less completely, as the vacuum is more or less perfect. If sufficient blood be not obtained at the first application of the cup or cups, they must be removed as soon as they cease to fill, the wounds wiped lightly with a sponge, and the glasses put on as before. The wounds generally cease bleeding when the cups have been taken off; and a piece of lint spread with ointment, or of sticking-plaster, is put on merely to keep them free from dirt. Such is the orthodox mode of performing this operation.

But if no cupping instruments are to be had, an off-hand cupping may be performed, with a little dexterity, with very simple materials. A small tumbler or teacup, either of which is preferable if bellying, or a tumbler of the old-fashioned shape known as a rummer, is the best of all—and answers for the cupping-glass; a bit of lighted tow or paper is then to be put into it, and as soon as the tumbler or cup gets pretty warm, and the air within is rarefied, it is to be turned down upon the skin, which will rise tolerably well into it. There is, to be sure, a little danger of slightly burning the skin: but as the tow is extinguished from want of air immediately the glass or cup is turned down, not much damage is done. The place of the scarificator. is supplied by wounding the skin, till it bleed,

ın half-a-dozen different places, within the circuit
of the glass or cup, with a lancet, or with the
point of a razor or of a sharp knife, after which
the glass or cup is put on as at first. This method
I was taught by an old campaigner, and it makes
a very good substitute for the regular mode.

It sometimes, though very rarely, happens
that, as with leeches, so in cupping, a little artery may be wounded, and bleed pertinaciously
in which case it must be treated in the same way
with needle and thread.

BLEEDING

Is commonly performed at the bend of the
elbow, and sometimes—though not frequently—
on the top of the foot. At the latter place it may
be performed without danger ; but at the elbow
there is always, to an unpractised person, danger,
of not merely opening a vein, but also of wounding an artery, which is a very serious and sometimes a fatal accident. I should, therefore,
earnestly advise *no one to attempt bleeding in
the arm without having been properly instructed*
by a proper person, or both he and the patient
will get into great trouble. But for the same
reason I would strongly urge all intelligent emi-
grants or travellers in out-of-the-way places,
and even clergymen, to learn how to perform

this operation, as it may be of the greatest service occasionally. Though it must be remembered that bleeding from the arm is not often, at present, so promptly performed after an accident as formerly, when it was the common practice, immediately after a person had become senseless by a fall or blow, to plunge a lancet into him as soon as possible, with the praiseworthy object of quickly restoring him, but with the almost certain result of hastening his journey to "that bourne from whence no traveller returns." It must, therefore, be well remembered, that, although I shall describe the modes of bleeding from the arm as well as the foot, I do, at the same time, urgently press on every unpractised person the danger himself and the patient both run, if he attempt bleeding in the arm, and therefore entreat him not to do so.

Bleeding at the bend of the elbow.

Along the arm, and upon its outside, runs a large vein from the root of the thumb up to the

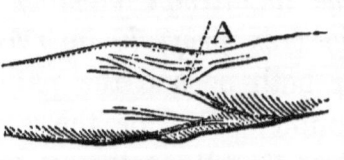

shoulder; and on its inner side another of equal size from the little finger into the arm above the elbow; a third vein of nearly equal size makes its appearance at the top of the fore-arm, just below the elbow, and very soon divides into a fork, one branch of which runs to the inner vein, and the other into the outer vein, just above the bend of the joint. The great artery of the arm runs close behind the inner branch of the middle vein ; and, therefore, this vein should not be bled in, for fear of wounding the artery behind it ; notwithstanding, some medical men do not hesitate to bleed in this, trusting to their superior skill— a piece of fool-hardiness that does not always escape unpunished.

The *outer branch of the middle vein* (A), which runs into the outer vein, is the *proper vein* to open ; but as the branching of the arteries about the elbow is very irregular, it not unfrequently happens that even behind this vein is a large artery : and therefore it should always be ascertained, by putting the point of the finger upon the part of the vein to be opened, whether any artery be there ; and if it be, it may be felt beating as a pulse. If no vein free from an artery can be found, then there is no choice ; but great caution is requisite to prevent mischief. A piece of broad tape or ribbon is now to be turned twice round the arm, a hand's breadth above the elbow, and its ends tied in a bow-knot, so that it may be easily loosened. The use of binding the arm thus is, to prevent the return of the blood, and make the veins swell

and jut well up. The operator now takes hold
of the fore-arm, applying the palm of his left
hand and fingers to it just below the elbow,
and passing his thumb over the outside, so
that its tip may lie upon the
vein to be opened, and by
slightly pressing prevent it rolling. The blade of the lancet is held by
the thumb and
finger of the
right hand, its
scales or co-
vering being
turned for-
wards, at an angle towards the tip, so as to
be out of the way. The other three fingers
of the right hand are then gathered together
and rested on or near the left thumb, so as to
form a support for the fore-finger and thumb
holding the lancet, the point of which, being
brought down to the skin, is made to pierce it
and the vein together, with a swinging motion
upwards, upon which the blood immediately
flows out. This swinging mode of using the
lancet is better than plunging it directly in, and
withdrawing it in the same direction, as it makes
a large opening, and the blood more freely flows
because the skin wound is necessarily larger

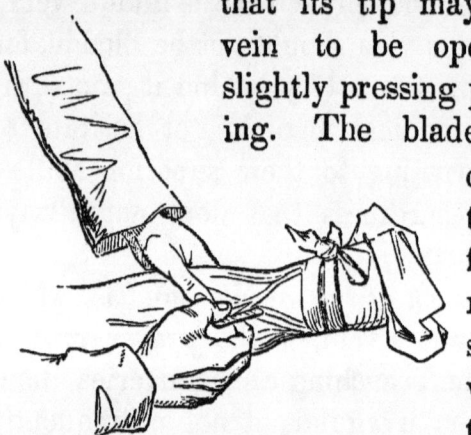

than that in the vein. It is a.so less likely to pierce through both sides of the vein, which, if done, is often followed by a large escape of blood beneath the skin, which swells up and stops the stream. The flow of blood from the vein is furthered by the patient grasping a stick and playing the fingers; for the former purpose, in ancient times when barbers practised bleeding, as they still do in many foreign countries, the pole, with which the front of barbers' shops may still be seen adorned in obscure streets and country towns, was held for this purpose; and the white stripe around it alludes to the tape or fillet : together they showed that " below for threepence he bled."

When the required quantity of blood has been taken, the thumb is put upon the wound, the tape untied, the blood on the arm cleared with a sponge, and a bit of lint or linen three or four times folded, and about an inch and a half square, is put on the wound, the edges of which are at the same time brought close together. This pad is to be fastened with a tape passed round the arm half-bent, in shape of a figure of eight, the cross being on the pad and the loops surrounding the arm. The tape should only be bound sufficiently tight to confine the pad; for if tighter, the blood cannot pass through the unwounded veins, which therefore swell, and the blood again bursts open the vein which had been closed. The arm should be kept quiet for two or three days, to give the wound in the vein time to heal, and avoid the chance of the blood again bursting out. If the person be timid, it is better to bleed him whilst lying down, otherwise he is likely to faint before the proper quantity of blood is obtained; but if, as is sometimes necessary, the person is bled to make him faint, he should stand up whilst the blood flows.

When a *vein is to be opened in the foot,* a garter must be tightly tied round the leg immediately below the knee; soon after which the veins on the top of the foot swell up; and if they do not the foot may be put into hot water, which soon makes them fill sufficiently. Of these the largest should be chosen, and it should be opened lengthways with a lancet, in the same way as in bleeding in the arm. When the quantity of blood desired has been obtained, the garter should be taken off; the person lie down at full length, and the wound closed with a pad of lint and a strip of sticking-plaster.

BLISTERING.

———

I HAVE already mentioned some of the milder modes of irritating the skin with mustard poultice, hartshorn and oil, mustard liniment, and croton oil; but it frequently happens that in inflammatory and other diseases of the body more active and speedy irritation of the skin is required, which is effected by blistering.

The extent of the surface to be blistered having been determined, a piece of common stickingplaster or strapping is to be cut out, of corresponding size, upon which blister-plaster is to be spread with the thumb (which should be continually wetted to prevent the plaster sticking to it) to the thickness of pretty thick brown paper, leaving, however, an edge or margin of the strapping about a third of an inch in width, for the purpose of fixing it upon the skin; to which the blister should be gently pressed, so that it may be completely applied and afterwards lightly bound on with a roller or handkerchief. In ordinary cases, twelve hours are sufficient for the part to become blistered; but if not by that time, the blister must be continued for twelve hours more. This, however, applies only to grown-up persons.

With regard to children, and very young children especially, a blister should not be continued more than three or four hours, by which time the skin has generally become smartly inflamed; and if the plaster be taken off, the blister usually rises a few hours after. The danger of leaving a blister on a child longer than the time mentioned is, that it oftentimes not merely raises the scarf-skin as a blister, but destroys the true skin beneath. The effort the constitution is required to make for the purpose of throwing off the slough thus formed, is sometimes greater than the child's strength can bear, and after lingering some days he dies of the blister instead of the disease which it was employed to relieve. If, however, the child escape this sad consequence, there is still the inconvenience of a slow restoration to health, much suffering, and a scar of size corresponding to that of the slough. So far as the unsightliness of the scar only is concerned, it is not matter of much consequence to a boy; but if a girl be thus scarred by a blister put on the upper part of the chest, on the throat, or at the back of the neck, it will be matter of bitter vexation to herself and her parents as she grows up; for even if the scar after many years subside, as it does occasionally, into a stain, the probabilities are, that she will carry it with her to her grave. Therefore *with children under ten years of age, never keep a blister on longer than till the skin has become well inflamed.*

Some persons are liable to a dreadful degree of inconvenience and suffering a few hours after the application of a blister, in consequence of being attacked with strangury, or a frequent desire, sometimes every ten minutes, to pass water, which is made only in very small quantities, often not more than a teaspoonful at a time, scalding hot, like boiling water, and producing extreme agony. This will continue for hours, and sometimes is accompanied with a few drops of blood.

All these horrible annoyances may, however, in general, be very easily prevented by the simple precaution of covering the blister-plaster with a piece of tissue-paper, which should be gently pressed with a warm finger till it become greasy throughout, by sucking up the fat in the plaster. Thus covered, the blister is to be put on as already directed.

If under any circumstances strangury come on, the blister must be immediately removed, and the part repeatedly dressed with spermaceti ointment till every particle of the Spanish fly, or of the grease impregnated with it, be removed. The person should also drink very freely of barley-water, or any other mucilaginous liquid free from acid.

It is always best, if there be opportunity of choosing time, to put on a blister immediately before going to bed, or making quiet for the night; as by so doing very many persons go to sleep and suffer little whilst the blister is drawing; and on waking in the morning only feel the part tight, and rather painful when endeavouring to change their position.

The pain and inconvenience attending the healing of a blister very much depend on the mode in which it is managed. If properly attended to, it is scarcely painful after twelve or sixteen hours, and will often be completely healed in forty-eight hours.

In *dressing a blister properly*, the principal object is to empty it of all the fluid it contains without breaking the skin. Before doing this, a piece of lint or linen, larger than the blistered surface, should be thickly spread with spermaceti ointment; two or three thick, small pads of lint or soft linen should be made; a sponge which has been soaked in water and wrung dry as possible, and a folded napkin, should be provided.

The lowest part of the blister is then to be snipped with a pair of scissors to the extent of a quarter or half an inch, so that the fluid may readily escape from it into the sponge, which should be placed immediately below the wound. As soon as so much fluid has escaped as will flow of itself, the remainder, wherever it bags, should be gently urged with the pad of lint towards the hole; but if it cannot all be made to escape there, one or two more small holes must be made with the scissors, and the remaining fluid, as far as possible, pressed through them. In emptying the blister, the greatest care must be taken not to tear or wrinkle the blistered skin : it should be made to lie as smoothly as possible, which it will do, except just at the edge, where it will of course double, on account of its previous distension by the fluid. The reason why the skin should be so carefully preserved unbroken is, that it is the best application to the inflamed and very sensible true skin beneath, defending it completely from the air, which would otherwise dry, and render it very sore, and also promoting the formation of the new scarf-skin from it.

The folded napkin is then to be laid over the dressing, and may either be fastened with a bandage, or left alone, according to the restlessness or quietude of the patient.

After three or four hours, this dressing may be removed; any fluid which has collected may be pressed out by the holes already made, or by making others; and fresh spermaceti dressing and napkin are to be put on as before.

The dressing should be changed again in four or five hours, and if any fluid remain it should be carefully pressed out; and this should be repeated so long as there is any fluid.

Generally, after the first three or four dressings, the inflammation set up by the blister-plaster ceases; the patient is at ease, so far as the blister is concerned, and it does not require dressing more than twice in the twenty-four hours till it has completely healed.

In some cases it is thought advantageous to check the quick healing of the blister; and then for the first two or three days it must be dressed with yellow basilicon; but after that time with spermaceti ointment.

In those untoward cases which will sometimes occur in persons of very irritable constitution, in whom sloughing of the skin to a greater or less extent follows the application of a blister, poultices

In those untoward cases which will sometimes occur in persons of very irritable constitution, in whom sloughing of the skin to a greater or less extent follows the application of a blister, poultices must be used, and the case managed in the same way as when the skin is destroyed by scalds and burns, to which the reader is referred.

VACCINATION.

VACCINATION, which has been declared to be one of the greatest blessings conferred on man, is now so freely offered to the poor, by the Poor-Law Commissioners, at the cheap expense to the parish of eighteen pence a head, that there is little chance of the Lady Bountiful of the village being disposed to set up as cow-poxer against the farrier, who practises inoculation for the small-pox at a shilling a head.

But to persons emigrating, it is of the utmost importance they should be able to enjoy protection, or at least the best protection possible for themselves and those dear to them, from that awful scourge, small-pox, which on its first appearance, in a newly settled or previously unvisited district, usually sweeps off hundreds, and has in America destroyed even whole tribes of Indians. Vaccination is so easily taught, so easily learned, and so easily practised, that trifling attention to a few simple rules may render any kind-hearted person the greatest benefactor to his neighbourhood, and therefore should never be slighted.

I shall take the liberty of quoting from my

F

friend DR. GEORGE GREGORY,* at present the
great authority on Cow-Pox, the principal circumstances to be attended to in conducting vaccination.

The introduction of vaccine lymph beneath the
skin, in consequence of which a peculiar vesicle
or little bladder is formed, and passes through
certain stages, till at last a scab drop off, con-
stitutes vaccination.

"The younger the lymph is," says GREGORY,
"the greater is its intensity. The lymph of a
fifth-day vesicle, when it can be obtained, never
fails. It is, however, equally powerful up to the
eighth day, at which time it is also most abundant. After the formation of areola, the true
specific matter of cow-pox becomes mixed with
variable proportions of serum, the result of common inflammation, and diluted lymph is always
less efficacious than the concentrated virus.
After the tenth day the lymph becomes mucilaginous, and scarcely fluid, in which state it is
not at all to be depended on. . . . . Infantile
lymph is more to be depended on than the lymph
obtained from adults." For the proper performance of vaccination, "Let the lancet be exceedingly sharp, and if fresh lymph is to be used,
its point must be introduced into the vesicle of
the child near at hand, in such way as to bring

---

* Lectures on the Eruptive Fevers. London, 1843, 8vo.

out upon it some lymph without drawing blood, and is then to be inserted into the arm of the child to be vaccinated." It should penetrate the corion* to a considerable depth. The notion that the subsequent effusion of blood will wash out the virus, and thus defeat our intention, is quite imaginary and groundless. Provided that a genuine lymph of due intensity has once come in contact with the absorbing surface of the true skin, the rest is immaterial. . . . In making the incision, the skin should be held perfectly tense between the fore-finger and thumb of the left hand. The lancet should be held in a slanting position, and the incision (puncture, rather) made from above downwards. . . . I would recommend that, with lymph of ordinary intensity, five vesicles should be raised, and that these should be at such distance from each other as not to become confluent (not to run together) in their advance to maturation. About the third day a blush appears distinctly at the vaccinated points; but GREGORY says, "by aid of the microscope the efflorescence surrounding the inflamed point will be distinctly perceived even on the *second* day. On the *fifth* day the cuticle (scarf-skin) is elevated into a pearl-coloured

---

* The true skin, which is the part exposed when the cuticle or scarf-skin, raised after a scald or by blistering, has been removed.

F 2

vesicle, containing a thin and perfectly trans
parent fluid in minute quantity. The shape of
the vesicle is circular or oval, according to the
mode of making the incision. On the *eighth* day
the vesicle is in its greatest perfection ; its margin
is tinged, and sensibly elevated above the surrounding skin. In colour the vesicle may be
yellowish or pearly. . . . The vesicle possesses
the umbilicated (indented) form belonging to
small-pox. . . . On the *eleventh* day the areola

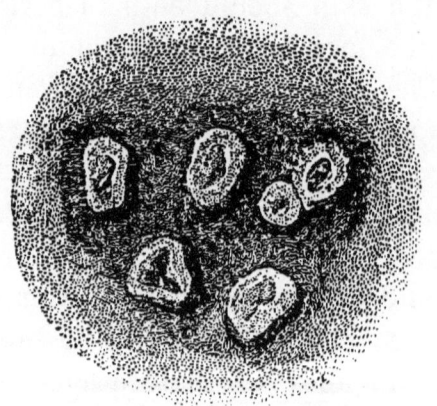

begins to fade, leaving in its decline two or
three concentric circles of a bluish tinge. Its
contents now become opaque, the vesicle itself
begins to dry up, and a scab forms, of a circular
shape and a brown or mahogany colour. By
degrees this hardens, and blackens, and at
length, between the eighteenth. and twenty-first
day, drops off, leaving behind it a cicatrix (scar)

of a form and size proportioned to the prior inflammation. *A perfect vaccine scar should be of
small size, circular, and marked with radiations
and indentations.* These show the character of
the primary inflammation, and attest that it had
not proceeded beyond the desirable degree of
intensity. Many of the most perfect scars disappear entirely as life advances. Until the
eighth day the constitution seldom sympathises.
At that period, however, it is usual to find the
infant somewhat restless and uneasy. The
bowels are disordered; the skin is hot; and the
night's rest is disturbed. These evidences of
constitutional sympathy continue for two or three
days. There is, however, much variety observ-
able here. Some children suffer lightly in their
general health throughout the whole course of
vaccination; others exhibit scarce any indication
of fever, although the areola be extensive and
the formation of lymph abundant."

In this way vaccination is to be managed when
the lymph can be obtained fresh from the vesicle
of a child who is passing through the disorder.
But it may happen that vaccination has to be
performed where no fresh lymph is to be obtained; and it may have to be procured from a
great distance, and much time may necessarily
have to pass by, as for instance, when required
for the Colonies, before it can be furnished.

To meet this emergency various modes of conveying lymph have been recommended. It has been collected in stoppled bottles, and in little glass bulbs, which will do well enough for two or three days; but GREGORY says, "if you attempt in this manner to transmit lymph to the East or West Indies, you will utterly fail." But he recommends "ivory points, which when well armed and carefully dried are very effective. They will retain their activity in this climate for many months, and they are found to be the most certain mode of sending lymph to our colonies." BRYCE of Edinburgh recommended vaccinating from scabs preserved in stoppled bottles, which might be sent to any distance; this, though a very excellent plan, requires more nicety than can be expected from unprofessional persons. It is also sometimes sent in charges, as they are called, between two small pieces of glass and hermetically sealed; but the points are best.

In vaccinating with a point, which is a piece of ivory, in shape like a very narrow lancet, the

proceeding is rather different from vaccinating with fresh matter. The point having been chosen, the dried lymph upon it must be moistened by breathing upon a few times. Punctures in the skin are to be made with a lancet in the same way as already directed, and then the point having been breathed on again, must be passed into each wound thus made, and gently pressed, so as to transfer the lymph from the point to the wounds.

It would seem scarcely necessary to mention, were it not that people frequently do not think it of consequence to pay any particular regard to it, that *the vesicle must be carefully guarded against being burst or injured;* for if it be, the progress of the disorder cannot be watched, nor its having passed through its proper course ensured.

To be able to draw a tooth moderately well, must be a great accomplishment to an emigrant settler in the backwoods, and is well worth acquiring. If he will submit to proper instruction before leaving civilized society, so much the better; but if not, with a few plain directions, and getting an old skull or two, from which the teeth have not dropped, to practise on, he may manage in a little time to draw teeth tolerably well; and if he break a tooth or two short off, or pull out a tooth and bring a bit of the jaw away with it when he begins to practise, he may console himself with the reflection that few doctors' apprentices have not done this in their early experience, and that the same misfortune occasionally happens to celebrated dentists, though neither one nor other talk about it. Besides, there are few persons suffering under the horrid pain of tooth-ache, more especially after experiencing it two or three times, who will not risk anything to be freed from their misery. It was formerly the practice to separate the gum from the tooth by digging a gum-fleam as deeply as possible round the tooth; but this is quite exploded now, and the strength of the wrist only is relied on.

The *front and the eye teeth are pulled out* with straight forceps, the blades of which are placed one behind and the other before the tooth, and the ends made to clip just before the tooth dips into the gum. The right hand then grasps the handles of the forceps ; whilst the fore-finger is at the same time thrust far in between them, to prevent too great pressure being made and the tooth snapped off. If it be an upper tooth, the operator steadies the patient's head by getting it beneath his left

arm, and then pulls down, giving the tooth a twist at the same time, by which it is soon drawn if the pull be steadily made. If it be a lower tooth, the operator steadies the head in the same way,

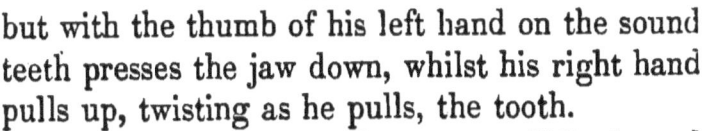

but with the thumb of his left hand on the sound teeth presses the jaw down, whilst his right hand pulls up, twisting as he pulls, the tooth.

Drawing a back tooth is a more difficult and complicated business, and is generally done with the instrument called a key ; the handle and stem of which are like a boot-hook, less the hook. The free end of the stem of the key has a deep solid lip, which is called the bolster, and on the top of this moves a strong shortly curved iron claw, which, by twisting the handle of the instrument, acts most powerfully, and drags the tooth out of its socket. Formerly it was the practice to cover the bolster with lint or leather, so as to prevent bruising the gum; but this is now given up, and the gum is left to take its chance. If an upper back tooth is to be drawn, the operator has most power and control, and can see best what he is about, if he set the patient on the floor, throw his head well back, and fix it between his own knees. If it be the lower tooth, he may place him in a chair. In either case the mouth must be held wide open.

The operator now introduces the key, with the claw thrown back, into the mouth, within the range of the teeth, places the bolster of the instrument against the gum of the tooth to be pulled out, then turns the claw across the top of the tooth, and lets it drop till it lock on the outside of the tooth just where it sinks into the gum. Here he steadies the claw with the forefinger of the left hand, and grasping the handle of the instrument, as he would the handle of a

corkscrew when about to pull out a cork, he
twists it from without inwards, and as he does this,

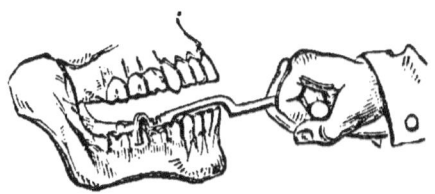

the claw acting as a lever, the fulcrum of which
is the bolster, lifts the tooth out of the socket.

Some prefer drawing the tooth outwards, in
which case the bolster must be placed on the
outside of the gum, and the claw made to catch
on the inside, after which the handle of the instrument is twisted outwards. The choice of
these two modes is rather matter of caprice than
anything else.

*Be sure in fixing the claw you grip the right
tooth, and take care it do not slip to the next*, or
a sound serviceable tooth may be drawn, and
the plague left behind—an accident which now
and then occurs, doubtless to the patient's great
satisfaction.

HOW TO PUT ON A ROLLER.

"Any body can do that," the reader is likely enough to say ; but let him try and roll his own leg, as he may do after a fashion truly, and then let him walk about a little and find how long it will remain without slipping about his heels. The fact is, rolling a limb seems a very easy business, but to roll it properly requires a good deal of knack and practice.

The common material for making rollers is moderately stout calico, which should always be washed before first using, to get out the alum and stiffening with which it is made up for the shops, but which prevent it lying close when employed as a roller. There is, however, a much better thing for those who can afford it, which is sold by CHURTON, in Oxford Street, and goes by the name of CHURTON's Roller. It is made of a sort of cotton-stocking material, and yields, so that many of the twists and turns requisite for applying a common calico roller properly are needless. For ordinary purposes, however, it is too expensive.

The breadth of a roller should be about that

of three fingers; its length must vary according to the size of the part to be rolled. The leg, which is the part most frequently requiring this treatment, needs a roller of six yards' length. If the thigh also have to be rolled, a second had better be tacked to the first, when it has been nearly used up, as a longer one than six yards is so cumbersome that it cannot be properly put on. The best time for putting on a roller, especially on the leg, is before leaving bed in the morning, as the veins, even under ordinary circumstances, are then less filled than after moving about for an hour, and the limb being consequently less large, the roller will then give it the greatest support, which is the object of its use.

A roller is generally rolled from one end only, and is then called a *single-headed roller* (*a*). But if it be rolled from both ends, so that the rolls meet in the middle, it is a *double-headed roller* (*b*); this is not, 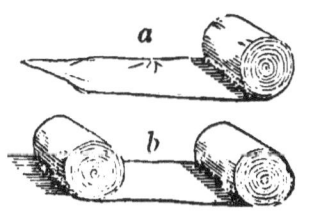 however, often used, as it is not very convenient.

For those who have the opportunity of having assistance, a roller can be better applied by an attendant than by themselves. But such as have not this good fortune or prefer being indepen dent, may soon learn to put on their roller very well, and without occupying much additional time in making their toilet.

It is not a matter of no consequence as to how much of the leg should be rolled, or at what part of it the rolling should be begun. A roller from the ankle to the knee is at least of only partial use, and may be very hurtful by making the foot swell. And if it be put on near the knee and rolled downwards, it is put on at a disadvantage, because it girts the veins and makes them swell below, and swell still more, the lower the bandage is carried, till at last they are rolled over, full as they are, and when by the continued pressure they have become equally emptied, the roller becomes slack and down it slips. In rolling the leg, therefore, the foot as well as the leg should be rolled, and the rolling should be begun at the toes and finished just below, or, what is much better, immediately above the knee.

In putting on a roller each turn round the limb should be made spirally, whence it is called *spiral rolling;* and each turn of the roller should overlap half of the turn just before made ; and so long as the limb is straight and of nearly equal size, this is easily done. But in passing from one part of a limb to another, especially where the joint forms an angle, as at the instep, in rolling from the foot to the leg, and higher up where the calf considerably increases the bulk of the leg, the roller will not fit without a little different management, which consists in turning the roller half down upon itself, or even more, slantways, which is called *reversing the roller*, till it will fit itself to the limb and half cover the previous turn as it had done at first. These slantways turnings-down of the roller often require three or four successive repetitions, till it get into proper trim to move on regularly. The knack of doing this is the great secret in applying a roller properly, and it is not one in a dozen people, professional or not, who can roll a limb well unless he have had pretty ample practice on his own person or that of others. This prologue concluded, let us now see—

How to roll a leg.

Presuming the person is right-handed, he will take a single-headed roller in his right hand, holding its circumference between his thumb and fingers, and laying its loose end on the top of the foot at the root of the toes, he fixes it there with the thumb of the left hand whilst the roller itself is carried beneath the sole and round the foot, and twice or thrice round in the same place till it get a hold on the foot. The roller is then to be turned round and round the foot towards the heel, each turn half covering the former one, and, as the roller passes beneath the foot, it is deli-

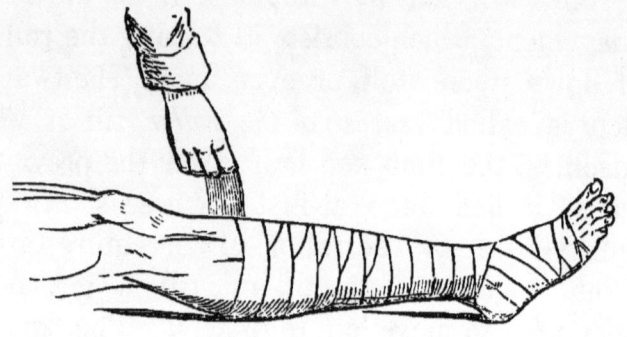

vered from the right to the left hand, and then as it passes over the foot, from the left to the right hand. Having arrived at the instep, the roller is now carried round the ankle, made to descend to the opposite side of the foot from which it had been brought, passed beneath the sole, and then carried round the ankle again; thus forming a figure of eight, one loop containing the foot, the other the ankle-joint, and the gripe of the loops on the instep; this figure-of-eight turn generally requires twice making to give the roller a good start up the leg. The roller is now turned round the leg, each turn half covering the former, and delivered from hand to hand alternately, from within to without or from without to within, according to which leg is being rolled. This spiral turning is to be continued so long as each turn will half cover the former but so soon as this

fails without one edge of the roller being left
loose and unfitting, which generally happens at
the spring of the calf, the roller must be turned
down on itself or reversed so that it will lie flat,
and then carried round the leg; at this part the
bandage often must be reversed several times,
and should always be reversed the same way,
before it can be again put on spirally, in which
manner it is to be continued up to the bend of the
knee, and if not taken farther, two or three circular turns one upon the other should be made,
and the roller fastened and cut off; or if all be
not expended, it may be taken lightly down the
leg again in the same way, till it be exhausted.
If, however, the roller be continued above the
knee, it will require to be passed behind the
joint around the thigh and down again behind
the joint to the leg, so as to make a figure of
eight, of which the gripe is in the ham, and this
requires to be repeated; after which the rest of
the roller may be turned circularly two or three
times round the thigh above the knee. By
making the figure of eight turn at this part, the
knee-cap is left uncovered, which is requisite;
for if it be rolled over, however carefully the
roller be applied and however many times it
may be reversed, and a great many times it
must be, to make it fit at all, directly the person
begins to walk about all that part of the roller

G

covering the knee slips, and as a necessary consequence the whole roller becomes loosened and
useless.

To roll the thigh

Is merely continuing to roll spirally from above
the knee to the groin, having reached which the
two or three last turns must be tacked together
and then a turn or two made round the hips, and
these tacked to the roller on the thigh so as to
prevent it slipping down, which without this it
certainly will do, as the thigh is bigger at the
upper than at the lower part.

To roll the fore-arm alone or the upper arm also

Is a very easy matter, it being scarcely ever necessary to reverse the roller. It is generally
only begun at the wrist and rolled upwards;
but if the fingers and hand become puffy and
uneasy, as they sometimes do, it will be. necessary to roll each finger separately with a narrow
bandage, and then roll the hand itself to the
wrist, after which the arm must be rolled as
directed.

The belly or chest is rolled

Generally with a flannel bandage as wide as two
hands' breadth and half a dozen yards long. The
roller is here put on spirally up and down till it
be exhausted. It is best to 'tack through the

first two or three rolls or turns before proceeding further, as otherwise the roller soon gets loose.

There are various other modifications of rolling, some of which will be by and by noticed; but for our present purpose all that is needful has been mentioned.

As an epilogue to rolling, two very useful bandages may here be described :—

THE T BANDAGE

Is the best contrivance for keeping a poultice on the seat, when such application is required for boils or abscesses in that neighbourhood, which too frequently end in fistula. It is also very useful to confine a poultice on the groin.

The T bandage is a very simple contrivance and very easily made. Its shape corresponds precisely to that of the letter T. That part of the bandage answering to the head of the letter forms a belt which ties round the belly immediately above the hips, and should be made of linen a hand's breadth 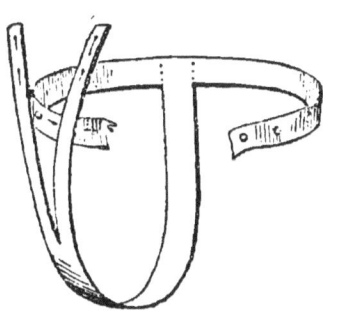 wide. The stem of the letter is formed by a

piece of linen double the width of the former and
sewn by one end to its middle, so that it lies
against the loins. This piece should be of sufficient length that it may be brought forwards and
upwards between the legs to the front of the
belt, over which its loose end is to be turned, and
being split a little way down, the two loose ends
thus made may be brought forwards and tied,
or it may be sewn to the belt without splitting.

If with this bandage a poultice have to be confined on the groin, the tail-piece must be inclined
to that side and fastened to the belt as may be
necessary.

THE MANY-TAILED BANDAGE

Is employed for the purpose of supporting a
limb, and it may be also for keeping on dressings
or poultices when the patient's state is such that
he cannot bear the fatigue of having his limb
rolled. This bandage is made of linen, and consists of one long band of roller width, across
which transverse pieces of the same width, but
of sufficient length for their ends to overlap each
other after surrounding the limb, are laid, one
half covering the other, and, thus placed, are
sewn at their middle to the long band.

To put on this bandage the limb is gently
raised of sufficient height to slip the bandage
beneath it, the longband being placed in corre-

spondence with the length of the limb; the ends of the cross-pieces are then pulled out and laid smoothly and regularly upon the bed, which done they are turned over the leg alternately, from below upwards, one over the other, till the whole limb is completely enveloped.

Strapping a limb with plaster

Is merely a many-tailed bandage spread with plaster, but without the circular straps being connected by a longitudinal strap.

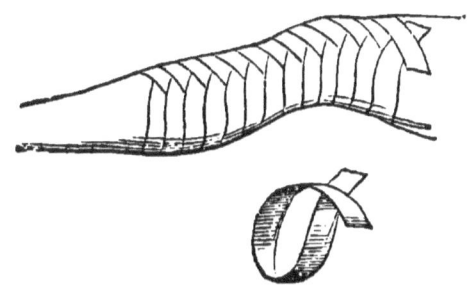

Having thus described the common modes of bandaging with rollers, I cannot avoid referring to

Bandaging with a kerchief,

which has been lately and advantageously brought into use at the suggestion and with the recommendation of the Prussian Surgeon Esmarch, Professor of Military Surgery at Kiel, in his essay, 'Kriegschirurgischen Technik,'* which ob-

---

* Translated by H. H. Clutton, Assistant Surgeon of St. Thomas's Hospital. Title: 'The Surgeon's Handbook,' by Professor Esmarch.

tained the first prize, given by the German Empress, at the Vienna Exhibition. Well worthy, indeed, is the book of all the merit and praise it has attained, and, in the author's own words, "to be employed as an assistance to the memory, which is accomplished better by illustrations than by words." With this notion, "his book accordingly contains many illustrations, with the shortest possible text." Among these is a cut of the several modes in which a half kerchief can be used as a bandage. It is printed upon kerchiefs, and these are served out to each soldier in the Prussian army. As the kerchief has become familiar to all who have had the opportunity and advantage of attending the Ambulance Classes which have been so ably carried on by Major Duncan almost throughout the length and breadth of our own land, I have not hesitated to reduce it as seen in the adjoining cut. Such printed kerchiefs can be obtained at any surgical instrument makers.

The kerchief should be of silk or linen, a square of a yard, or a yard and a quarter ; and it should be divided into halves by cutting across cornerwise.

The reference to the cut shows the mode in which each half kerchief appears when applied, and it will not require any great cleverness to put it on in like manner.

*

OF LANCING THE GUMS.

YOUNG children whilst cutting their first set of teeth often suffer severe constitutional disturbance. At first there is restlessness and peevishness, with slight fever; but not unfrequently these are followed by convulsion-fits, as they are commonly called, which depend on the brain becoming irritated, and sometimes under this condition the child is either cut off suddenly, or the foundation of serious mischief to the brain is laid.

Sometimes only one child in a family will be subject to fits during teething; but at other times every child is attacked, and the elder CLINE used to mention an instance in which several children of one family, as they arrived at this age, were successively attacked with convulsions and destroyed by them. The remedy, or rather the safeguard, against these frightful consequences is trifling, safe, and almost certain, and consists merely in lancing the gum covering the tooth which is making its way through.

When teething is about, it may be known by the spittle constantly drivelling from the mouth, and wetting the frock. The child has his fingers constantly in his mouth, and is continually biting any hard substance he can get hold of. If the gums be carefully looked at, the part, where the tooth is pressing up, is swollen and redder than usual ; and if the finger be pressed on it, the child shrinks and cries, showing that the gum is tender.

When these symptoms occur, the gum should be lanced, and sometimes the tooth comes through the next day, if near the surface ; but if not so far advanced, the cut heals, and a scar forms which is thought by some objectionable, as rendering the passage of the tooth more difficult. This, however, is untrue, for the scar will give way much more easily than the uncut gum. If the tooth do not come through after two or three days, the lancing may be repeated ; and this is more especially needed if the child be very fractious and seem in much pain.

Lancing the gums is further advantageous, because it empties the inflamed part of its blood, and so relieves the pain and inflammation.

Lancing the gum is very easily managed, and any intelligent person, after seeing it done once or twice, will do it very effectually. CLINE taught the mother of the family already mentioned to do this ; and after lancing her children's gums she never lost another, at least from that cause ; for, so soon as the teething symptoms appeared, she looked for the inflamed gum, lanced it, and they ceased.

The operation is performed with a gum-fleam, the edge of which must be placed vertically on the top of the inflamed gum and moved along, pressing firmly at the same time till the edge of the fleam grate on the tooth, and the business is finished. The relief the child experiences in the course of two or three hours is often very remarkable, and he becomes lively and cheerful.

SWOLLEN VEINS.

LABOURING people, especially women, and often those who are careless of their bowels, allowing them to be continually costive, are liable to have the veins of one or both legs become very large and swollen. If left alone, the whole leg, and sometimes the thigh up to the groin, become more or less completely covered with a network of these swelled vessels, which run in wavy directions and sometimes are seen to cross each other. After long standing about and towards evening the veins become enormously swollen, render the limb heavy and painful, and walking wearisome and difficult. Very commonly the skin inflames on some part of the leg, and an ulcer forms which occasionally bleeds, and is always very difficult to be cured. Sometimes, without any ulcer, one of the veins suddenly bursts, and the person loses a large quantity of blood, and naturally becomes much alarmed.

Treatment.

Very rarely can swollen veins be cured, if they have existed for a length of time. All that can be done is to prevent their greater enlargement, and even this cannot always be prevented.

It is absolutely necessary that the bowels
be properly moved every day, so that there
should not be any lodgment in them, the pressure of which prevents the free passage of the
blood from the limb. Five grains of rufous pill
every night, and three grains of blue pill occasionally, answer the purpose very well. Some
persons, indeed, are able to keep their bowels
regular by merely drinking a tumbler of cold
water directly after getting out of bed and before
taking any food. This being attended to, the
limb must be rolled, as already directed, with a
long roller of at least six yards' length, beginning
from the toes and continued above the knee, if
the veins of the leg only be enlarged ; but if those
of the thigh be also, the roller must be continued
up to the groin, and two or three turns made
round the hips, to prevent it slipping down. The
roller should always be put on before getting out
of bed, as the veins being then but partly filled,
the roller affords most support; but if the limb
be not rolled till some time after the person has
been up and about, it is comparatively of little
use.

Sometimes, instead of a roller, a *laced-stocking*
is used. This consists of a long jean gaiter with a
foot, fitted to the limb as high as may be required,
but which, instead of buttoning, is open along its
whole length, each edge furnished with eyelet-

holes, and kept together with a lace passed through them from one to the other, and from the foot upwards. This stocking also must be put on before the person leave his bed. When the stocking fits, which it rarely does once in twenty times, it is certainly much better than the roller, as it makes and preserves more equal pressure, and is less likely to slip. But it is an expensive article, and not often within the means of the poor to pay for ; and a roller, with a little practice, may be properly put on, and serves the purpose required very well.

When an enlarged vein bursts, the blood sometimes spirts out smartly, and a good quantity is very quickly lost. It may, however, be immediately stopped by putting the finger on the bleeding part, and laying the person down flat, either on the ground or on a bed. A little pad of lint is then to be put on, and bound fast with a roller, which should first be applied upon the foot and rolled up over the pad and above the knee, or higher, according to circumstances. The person should be kept lying in bed for a few days, in which time the wound heals, and the pad may then be removed, having first soaked it for a few hours in a wet poultice. A small piece of plaster may afterwards be put on, and the leg rolled as before.

BRUISES

ARE very common, and often very troublesome accidents, in consequence of either a heavy weight falling upon some part of the body, or the person falling from some height, and heavily. At first the part swells, and then blackens, with the blood which escapes beneath the skin from the small vessels burst by the blow. After a day or two, or more, according to the severity and extent of the bruise, the colour changes to dirty green, and the skin around the bruise has a greenish-yellow hue. By degrees the colour changes to all sorts of shades between black, blue, green and yellow, and then begins to subside, till the skin recover its natural colour more or less quickly. Occasionally, when much blood has been thrown out, it is not thus removed, but remaining, forms an imperfect abscess, which at last bursts through the skin, and is often troublesome to cure.

Treatment.

The best application for a bruise, be it large or small, is moist warmth; therefore a warm bread-and-water poultice, or hot 'moist flannels, should be put on, as they supple the skin, so that it yields to the pressure of the blood beneath, and thereby the pain is lessened. If the bruise be severe, and in the neighbourhood of a joint, it is well to apply some leeches on grownup persons, but not on young children, for the reasons already mentioned in speaking of bloodletting by leeches. Ten or a dozen should be spread over the whole bruised surface, and afterwards a poultice, or warm flannels, applied. The poulticing or fomenting should be continued so long as the pain and swelling remain; and it may sometimes be necessary to put on the leeches a second and even a third time. If the bruise be on a joint, the poulticing will often require to be continued longer, on account of the stiffness which usually remains for some time; and when left off, it is well to wrap up the joint in a soap-plaster. If the bruised part be the knee or the ankle, walking should not be attempted till it can be performed without pain, and at first should not be persisted in but for a very short time, and not to fatigue the part. Inattention to this point often lays the foundation for serious mischief in these joints, and in scrofulous persons, especially those who are young, may run on to the loss of a limb.

"*A black eye*" is generally no more than a bruise of the eyelids, spreading more or less

over the face, according to the size of the instrument by which it is inflicted. The greater
number of persons who get a black eye deserve
it, and, so far as I am aware, there is no remedy
save warm bathing, which will hasten its removal ;
but it is often a very tedious business. The
only thing to be borne in mind is not to get a
black eye; if you do, you must be content to
bear the disgrace for a few days, if you deserve
it. But if it have been an accident, there is nothing to be ashamed of, and a small draught of
patience will be a sovereign remedy.

Shutting a door or a drawer upon a finger. This
is one of the most severe forms of bruises, as the
door or drawer being usually closed with violence,
the finger end is sadly jammed, in general without but occasionally with a wound of the skin
or tearing the nail half off. Few persons have
escaped this accident, and it is therefore, almost
needless to mention how excruciating is the agony
for a few minutes. If the end only of the finger
be nipped, the nail very soon blackens, in consequence of the blood escaping from the broken
small vessels, and, being pent up beneath the
unyielding nail, which it separates from its attachment beneath, causes pain for a few days,
till the sensitive parts become accustomed to its
intrusion. But the detached part of the nail dies,
and, according to the extent of the mischief, even

the whole nail may die, and is replaced by a new portion of nail, or an entire new nail, which pushes underneath it from the root at the quick of the nail, and gradually but very slowly thrusts it on to the tip of the finger till completely loosened and thrown off. In general this process, after the first few days' pain, runs on without much inconvenience, but sometimes the injury is so great that violent inflammation of the nail joint of the finger comes on accompanied with severe pain, matter is formed, and the whole nail is quickly thrown off; which done, the tender skin under the nail soon hardens and has an ugly appearance, but the new nail puts out from the root, and after some weeks again ornaments the finger.

Treatment. The most speedy mode of procuring relief immediately after the occurrence of this accident is to plunge the finger into as hot water as can be borne. By so doing the nail is softened and yields or accommodates itself to the blood poured out beneath it, so that the agony is soon diminished. The finger may then be advantageously wrapped up in a bread and water poultice. On the following or on the third day, the blood has clotted, and, separating into its clot and fluid parts, the pressure it makes on the sensible skin under the nail may be relieved by scraping the nail with a penknife or piece of glass till it be so thin that the scraping give a sharp pain

H

from its nearness to the sensible skin ; the remaining thin nail then bulges and the pressure on the sensible skin is thereby relieved. But if the squeezed part of the nail be very black, and if it be very tender when touched, then it is best after scraping to make very carefully with a penknife a small nick through the still remaining nail over the black blood, and immediately it is cut through the watery part of the blood oozes out, the pressure almost entirely ceases, and almost instantaneous relief follows, but it rarely prevents the nail being thrown off.

If all parts of the end of the finger be injured, then are produced nearly the same results as from an aggravated whitlow (which see), and the whole, bone as well as soft parts, may mortify, and be thrown off or require amputation. This is not of rare occurrence in persons of unhealthy constitution, and therefore a jammed finger is not a thing to be lightly esteemed.

What has been said regarding the finger ends applies also to jammed toes, which are usually produced by the fall of heavy weights upon them.

WOUNDS.

A PRETTY long list, including their complications, do wounds present.—Cuts, bruised cuts, tears or rents, scratches and pricks of common language ; incised, contused, lacerated, and punctured wounds of the doctors. All these are varied by the violence with which they are inflicted, and by the sharpness, bluntness, cleanliness, or dirtiness, of the instrument which produces them, and materially influences the quickness or slowness with which the wound heals.

A *clean cut*, as it is called, may be produced by any sharp instrument, as a knife or scythe, the edge of which is drawn quickly across any part; but when made by the edge of a chisel, axe, or reaping-hook, the edge is forced in suddenly with some little force, and causes the slightest possible form of bruised wound; in common language, the part is said to have been *chopped into,* or *chopped off;* but it is scarcely more serious than a clean cut. A *bruised wound* is produced by the edge of a heavy piece of timber or stone falling on a part. *Stabs* are much the same kind of wounds, if the instrument be sharp and clean, and have not injured, as it too frequently does, important deep-seated parts among which it has penetrated.

These, especially cuts and chops, if they be
not large, and have not divided any important
parts, are the most simple and manageable
A large piece of flesh, even a finger or toe, may
be cut or chopped off, yet the wound generally
heals kindly and readily under proper treatment.
But occasionally a very trifling and simple cut, in
an irritable or intemperate person, will prove fatal,
and sometimes even where this constitutional
disposition does not seem present ; a sad example
of which occurred some years ago in a nobleman
who chopped off his toe with an axe, and died
soon after with locked-jaw.

If *a cut or chop* be not very deep, and if it do
not bleed much, or even if it do bleed, but the
bleeding can be stanched by bathing for a short
time with cold water, it is generally of not much
consequence, and can be easily and simply treated.
The corresponding edges of the wound should be
brought together as perfectly as possible, and
while thus held, some strips of plaster must be
laid across the wound, with small spaces between
every two, so as to allow the escape of an oozing
fluid, which often continues for some hours. The
edges of the wound should not be dragged tightly
together, but merely kept in place by the plaster ;
and if the wound be in the finger, arm, toe, or
leg, it is better that the ends of the plaster should
not overlap, as there is always a disposition to
swell in the neighbourhood of the wounded part ;

and then, if he ends of the plaster do overlap, it
forms a tight band, which, at the least, causes
unnecessary pain, even if it do not set up, as it
sometimes does, greater mischief.

If common sticking-plaster, or strapping, as it
is generally called, be not at hand, various simple
substitutes may be employed. Court-plaster, in
strips, may be laid across the wound. A coarser
proceeding, but of the same kind, is, drawing
out some thin bands of tow, winding them lightly
round the part, and then smearing with gum-
water till the tow stick. I have often, when a
boy, had my cut fingers treated and quickly cured
in this way. Isinglass-plaster may also be used.
White of egg, smeared on linen, will also answer
the same purpose. If, however, none of these
materials be at hand, we can manage pretty well
by carefully winding a bit of soft linen round
the part; the oozing from the wound quickly
moistens it, and, as it dries, the linen sticks firmly
together.

Whenever a part has been completely swathed
with either of these dressings, it will be necessary
to watch whether it become painful by the swelling rendering the bandage tight. If it do
become painful, the bandage should not be at
once taken off, but the blade of a pair of scissors
should be carefully run beneath it, and divide it
from end to end on the opposite side to the wound.
This generally gives instant relief, and the dress-

ing requires no further meddling with. It is always better to leave the dressing, as long as it remains fast and without pain, usually three, four, or five days; and if then taken off, either the whole or the greater part of the wound will be found united.

But if the wound or its neighbourhood become painful, and throb, it may be presumed either that the dressing do not agree, and irritate, or that matter be forming, or have formed, and not being able to escape, causes the disturbance. In either case the dressing must be removed, which is best done by soaking in water of an agreeable warmth, or by covering with a wet poultice, which after a few hours softens the dressing, so that it can be easily removed. If the wound be not inflamed, that is, red and tender, and if the discharge from it be good, that is, straw-coloured and of a creamy consistence, having a fair resemblance to thin boiled custard, the dressing may be re-applied; but if the edges be red and inflamed, or if they be pale and flabby—if the wound gape, if the matter be watery and stinking, then a single strap or band is to be applied to keep the edges near together, and the part must be covered with a poultice till the pain and inflammation cease, and the matter be of a good kind.

Messing a wound of this kind with friars' balsam, tincture of benjamin, or any such filth, is to be utterly eschewed, as they will only hinder instead of encouraging the union of the wound;

and for the same reason greasy applications
should be avoided.

A clean stab,

If of depth, although not causing any serious
mischief, generally does not heal so readily as a
cut, because it often unites near the surface,
whilst its bottom lodges matter; and therefore,
although for some days it may seem to heal very
steadily, yet then it becomes painful, the wound
opens, a gush of matter follows, and this may be
repeated once or twice before the cure is completed.

A bruised cut

May be made by a sharp-edged heavy piece
of wood or iron falling on a part. It is well at
first to try to unite the edges by sticking-plaster,
as in a clean cut; but if there be much bruising, this rarely succeeds; the bruised part is
more or less extensively killed, and must slough,
as surgeons call it, or form a core, as it is
vulgarly said to do, which slough or core has
the appearance of a piece of wetted buff-leather,
and is of an ashy colour; it must separate and
be thrown off before the wound can heal. In
such cases it is best to apply a bread-and-water
poultice first, to moderate the inflammation set
up; and as soon as matter begins to form, and
the extent of the slough is marked, which is

shown by the dropping in of the dead part, and
a narrow raw line between it and the living, then
the bread must be changed for a linseed-meal
poultice, which should be continued not only till the
slough have come out, but till the gap is filled up
by *new flesh*, or *granulation*, as it is called. When
the new flesh gets above the edges of the wound,
it is commonly known by the name of *proud flesh*,
and wrongly supposed to prevent the healing of
the wound, for it is in reality the material which
is produced for that special purpose.

When the hollow of the wound has thus filled,
the poultice may be left off and the sore lightly
bound with straps of sticking-plaster; or a linen
bandage moistened with cold water, and carried
a few turns round, will often answer the same
purpose. If the new flesh rise much above the
wound, it must either be kept down by pressure,
or it may be brushed lightly over with a bit of
blue-stone or blue vitriol (sulphate of copper);
and sometimes merely a piece of dry lint will suf-
fice. If neither of these be employed, a sort
of cauliflower of proud flesh is formed, which
prevents the skin shooting from the edges of the
wound over the sore, and hinders the cure.

A torn or rent wound,

Generally caused by a hook or nail, is one of the
worst kind, more especially if much skin alone be
stripped from the flesh beneath, as a large portion,

or even the whole of it, generally dies, and is thrown off, leaving a large sore very difficult to heal.

If the skin be merely torn, without being stripped, the torn edges may be lightly brought together with a piece of plaster, and a poultice applied. But if stripped up also, then, after gently washing with warm water, the skin should be laid down on its place as near as possible, a single strip of plaster put across to confine it, and then the whole covered with a bread-and-water poultice. The poulticing must in either case be continued till the slough of the torn edge or of the larger piece of skin have separated, and till the new flesh have formed, after which the wound must be treated as a common sore, with poultice or dressing as best suits.

Wounds by Gunpowder.

Another and very severe form of torn wound, which is accompanied also with much bruising, is that produced, frequently on the hand, by the accidental discharge of a fowling-piece, or by the blowing up of a powder-flask in charging a gun. In this accident the person may be thankful if he only blow off a finger or even the whole hand, for there are not wanting instances in which people have been shot dead on the spot. Such injuries rarely bleed much, unless some very large artery have been shot through, and not always if then. The person's immediate safety is to be provided

for, according to the directions given in reference
to "Bleeding from Wounds." If medical aid
can be procured, do not delay to obtain it as
quickly as possible, for the case is most serious.

I have adverted to this subject, to give a word
of prudential advice to those " who go a gunning,"
as the Americans call it, rather than to remedy
the mischief when done. First, in regard to the accidental discharge of a gun. By this I do not mean
to refer to all the mischances which may originate
from a number of persons occupying themselves
on the same business in a small space, only sufficient for one or at most for two, with safety, so
that it is rather the protection of a kind Providence, than the caution of his companions, that
one or more of the party be not shot, especially
if shooting on different sides of the same hedge or
in high cover. But I wish to warn sportsmen of
the mischief they may do themselves from their
own carelessness with their own gun. Charles
Alken's caricatures are truthful representations
of the results of carrying a gun cocked, and of
carrying it so that the trigger or the hammer of
the lock getting entangled either in the sportsman's
dress, or in going through cover, discharges the
gun, and often more or less severely wounds the
bearer, or if he have a companion gives him the
full benefit of the charge. I recollect many years
ago being out with a party of young people, and
in our walk we came to a ditch and high bank :

some of the party sprung over the ditch and
crested the bank, but before they had cleared it,
another who was behind, in making his jump,
caught the trigger of his gun in his shooting-coat ;
the gun went off immediately, and the charge
whistled so close by the side of one of the forward
party that for the moment we thought he must
have been shot. We were thankful, however,
to find he had escaped. Therefore, *never carry
your gun cocked*—if you know anything about the
matter, there is always time enough to cock it
and bring down your game, if you can.

Second. As to loading. It is always best to
have a loose charger to your powder-flask instead
of loading from the flask itself, as if there be any
explosion the mischief in that case will be trifling,
whereas in the other it is generally very serious.
A charge of powder may explode as it passes
down the gun-barrel, if by chance a bit of lighted
wadding remain in it from the last discharge.
More commonly, however, it is caused by a bit
of tow which had been used in cleaning the gun,
having been left by a careless servant in the
chamber and becoming lighted by the first or
second discharge. On one occasion, I myself
barely escaped this accident. I had discharged
my gun once, and, having loaded again, made two
attempts to fire without effect, the cap snapping
on the nipple and no discharge following. I
therefore drew the whole charge, and examining

the chamber with the screw, drew up a lump of tow, which the servant had negligently left when professing to clean the gun. Had this lighted when the gun was discharged, as was most probable, the powder, as it entered the chamber at the loading, would scarcely have failed exploding and blowing off my hand. Therefore *if you would be safe, clean your own gun, and previous to loading again, examine it with the rod and see the chamber is clear.*

I may also take the opportunity of exhorting the sportsman not to be so eager in his sport as to leave the ramrod in his gun, the consequence of which is that if the charge of shot be down, the gun bursts when fired. And the same accident is likely to occur, if by stumbling the muzzle of the fowling-piece is thrust into the ground and gets an additional charge of earth not fitting for it.

These observations, I trust, will not be considered out of place, for though not exactly " Hints on Emergencies," they are "Premonitions or Forethoughts " not to be slighted, although too frequently little remembered till the mischief is done.

Scratches

Are shallow rents not penetrating through the skin, and are commonly unheeded, as not requiring attention ; but if irritated by soap, pearl-

ash, or filth of any kind, they often become of
serious consequence, and sometimes even fatal.
Trifling, therefore, as scratches seem, they ought
not to be neglected, but should be covered and
protected, and kept clean and dry till they have
completely healed. If however, on the contrary,
they be irritated by wet and filth of any kind,
they not unfrequently inflame, becoming poisoned,
as it is vulgarly called. The inflammation
spreads, the neighbouring parts swell, and, unless
properly treated, the limb or the life may be lost.

Under these circumstances, no time should be
lost in applying to the doctor, if such can be had ;
the part in the meanwhile being wrapped up either
in a large bread-and-water poultice or in hot
flannels repeatedly applied, and leeches in good
number may be put on at some distance from
each other. But if neither leeches nor the doctor
can be had, the skin, where most red and swollen,
may be cut through till the blood come, but not
deeper, for fear of doing mischief, with a sharp
knife or razor, here and there, to the extent of
an inch, or an inch and a half. It is not proper
for any one to make these cuts, except when no
medical man can be had ; but if there be none,
it is justifiable, and valuable lives might be saved
by cautiously proceeding in this way.

Usually, after inflammation thus produced,
either large cores are separated, of the parts beneath the skin, which make their way out like a

common abscess ; or the skin itself also sloughs.
Continued poulticing is requisite till all the
sloughs come away and the wounds have
healed.

Pricks or punctured Wounds,

Though the most trivial in appearance, are often
the most serious in their consequences. They are
commonly produced by running in a splinter or a
thorn, and are rarely followed by more than a drop
or two of blood, and sometimes there is not even
any bleeding. The person pulls out the intruder
if he can, or if he cannot, leaves it to work out, as
he calls it, which it sometimes does after being
painful for two or three days, and matter forming
round it. But occasionally, whether the splinter
or thorn be pulled out or not at first, very serious
inflammation is set up in the part, and symptoms
of a slight degree of locked-jaw may come on ;
or the person may be destroyed by irritative fever
without or with well-marked locked-jaw. I have
known instances where the jaw-muscles were
stiffened, in one instance for an hour and a half,
after a thorn had been run into the finger and
been pulled out ; and in another locked-jaw
continued several days, after the running in of
a splinter which also had been removed at once.
And some years since a medical man died of
locked-jaw not many hours after having pricked
himself with a thorn, whilst out shooting.

If the splinter or thorn can be easily got out at once, it certainly should be. But there is often much more serious injury done by poking after it with forceps or the point of a knife or needle, and squeezing the part violently, than if it were left alone. If it be determined to have it out at once, and it cannot be readily got at, it is much better to make a cut, with a knife or lancet, along the course the splinter or thorn seems to have taken, so as more completely to expose and better to get hold of it. But even then much poking and squeezing should not be persisted in, as matters are thereby made worse; whilst, on the contrary, the wound has been improved by converting it from a prick into a cut, by making a more ready escape for the splinter and for any matter that may form, and thus lessening the probability of constitutional excitement.

If matters go on untowardly, the finger or toe, which is generally the part wounded, becomes very painful and tender, throbs violently and swells. The pain runs up the limb to a greater or less extent, and either the swelling spreads along it; or one, two, or more red thread-like lines are seen running from the wound towards the trunk; or there are both the swelling and the red lines.

These are serious symptoms and require prompt attention; leeches in good number must

be applied, it may be again and again, in the
neighbourhood of the wound, and the part wrapped
up in a large bread-and-water poultice, or hot
moist flannels, so as, if possible, to lessen the inflammation : and a smart dose of calomel, with
castor oil a few hours after, should be given.
Sometimes the wound itself is very tender, and
should then at once be cut into deeply, if on the
finger or toe. At other times it is neither tender
nor distinguishable, but some part a little distant
from it is very tender : if so, if it be swollen and
very red, and feel as if containing matter, it should
be cut into. This proceeding, whether there be
matter or not, is often followed by relief and the
inflammation gradually subsides, the wound occasionally, though not always, yielding matter.

The principal cause of all the mischief, when
the finger or hand, the toe or foot, has been
pricked, is, that the skin is so thick it will not
yield, when the sensible skin or parts beneath it,
having been wounded, inflame and swell; consequently, the inflamed part is, as it were, tightly
bound up, or squeezed as in a vice by the hard
skin. This readily explains the relief that almost
immediately follows cutting through the skin, as
thereby the inflamed part is at once relieved from
the squeezing; and any confined matter readily
escapes, instead of being compelled to burrow
along till it can find some place where, the skin
being thinner, it may burst through.

If either of the wounds already mentioned have been made by anything foul, as a dirty or rusty knife, a dirty piece of wood, horn, or the like ; or if the wound have been made by falling upon gravel, in which case the dirt can rarely be completely washed out,—it is better not to attempt uniting it with plaster or bandage at first, as it almost invariably festers. But a poultice should be immediately applied, and continued till inflammation cease and a good discharge be established ; after which it may be gently bound up with sticking-plaster or linen as already directed.

Those who " love to go a angling " occasionally make a catch they did not intend by hooking themselves, whilst incautiously holding the hook between the lips or fingers, and jerking the line by treading on or tangling it upon a bush or on the boat's side, as may be. Persons unskilled in such matters think it proper to wriggle the hook about, and then pull it out, as they would from a fish's mouth. This however is as unfitting as it is painful mode of treatment, for the barb of the hook, manage as well as you may, cannot be freed from the flesh, and can only be pulled out by dragging away some of the soft parts in which it is tangled. The best, readiest, and least painful mode of managing this accident is, first to grasp the stem of the hook tightly, and with a sharp knife rip off the line and clear the stem

I

of the binding silk, then make no attempt to withdraw the hook by the wound through which it has entered, but press the blunt end downwards, so that the point should be made to travel onward till it penetrate the skin and free the barbed point, which is then to be taken hold of and drawn further out in such way that the remainder of the hook follows through the last-made wound. Rarely any inconvenience beyond a few hours' smarting follows the accident, if thus managed; but if the finger be painful, put on a poultice.

ANIMAL-POISON WOUNDS.

POISONOUS Animals of any great severity, with the exception of the adder, we may be thankful this country is exempt from. There is a vulgar notion that the toad is poisonous, which is erroneous: when handled roughly, however, the secretion of some of the glands of the skin sometimes spirts forth; and, acccording to Dr. JOHN DAVY, if it fall on the face, produces a smarting sensation. Lice, bugs, and fleas—which, besides their ordinary smart, in some irritable persons occasionally produce a swelling which, if the bite be on the eyelid, especially of children, will now and then close the eye—are not poisonous; but merely produce inconvenience by the introduction of their sucking organs.

The harvest-bug,

Which is a very minute tick of a brilliant crimson colour, though not poisonous, is a terrible plague in autumn, as it buries itself very quickly in the skin of the legs of persons walking in the stubble. It produces intolerable itching; and if, instead of being carefully picked out, it be crushed and left in the leg, keeps up irritation, and if much scratched, troublesome sores will follow which often remain for weeks. KALM speaks of the American tick, which is another plague of the same kind, but much worse. It is very common in the woods of North America, and one cannot sit down without being quickly covered by them. They do not give much pain till they have half buried themselves in the skin, but then produce violent itching. At first they are so small as scarcely to be seen; but as they continue sucking, become as big as the end of the finger. When once thus buried there is difficulty in getting them away, as rather than leave their hold they allow themselves to be torn in half, and their trunk remaining in the wound, very troublesome sores follow.

The Chigoe

Is another of these burrowing animals. It is a kind of flea very common in the West Indies, and buries itself, in the soles of the feet, or about the toe-nails, for the purpose of depositing its eggs in

a little round vesicle, which sometimes, before it
is noticed, increases to the size of a small pea,
and is of a bluish colour. The bag requires to
be taken out with the point of a needle, without
piercing it, by separating it from the skin quite
round and drawing it out. If any of the bag be
left behind it causes violent inflammation, and
occasionally mortification, of the toes.

Gnat-bites.

Although gnats are sucking insects, yet, according to KIRBY and SPENCE, they "instil into
the wound made with their mouth a poison, the
principal use of which is to render the blood more
fluid and fit for suction." The grievous smart,
which these insects' bites produce, most persons
have experienced ; and not unfrequently, if the
eyelid be bitten during the night, it swells so
much that the eye can scarcely be opened in the
morning. The *Mosquito* is another species of
the same kind, and is still more bloodthirsty
than the gnat.

Smearing the part with olive-oil is the best
remedy for these bites.

The Stings of Hornets, Wasps, and Bees

Are the result of wounds not made by the insect
to obtain food, but in anger and for its own de-
fence ; and into these wounds, which are made
by a sharp dart at the extremity of the body, and

which is hollow, poison, secreted in twisted tubes (*a a*) which pour it into a little bag (*b*) specially formed as a reservoir, is thrown by the sting dart (*d*), which protrudes from between a sheath formed by two side plates (*c c*) at the end of the insect's body.

The agonizing pain which follows the sting of a hornet, wasp, or bee, more especially that of the former, is probably, for the time, as great as any to which we are liable. The throbbing is intense, and if the skin be thin, and loosely connected with the parts beneath, a swelling quickly rises; the skin becomes tight, shiny, and almost transparent, as if air or water had been forced beneath it. This swelling comes on very rapidly and spreads very quickly; so that if the eyelid be stung, the eye is quickly closed, or, if the lip, it becomes twice or thrice its usual size. Not unfrequently, if the insect have been irritated, it stings with such good-will that it leaves a great part, if not the whole, of its stinging organ sticking like a dart in the wound; and though it escape, thereby destroys itself. The hornet's sting is the most severe, and the bee's least so; but if a person be stung by many bees, he suffers severely; and there is an instance on record in which a man, attacked and stung by a large number cf bees on the chest and throat, died in the course of ten minutes after being thus injured.

It is always necessary to look carefully and see if any part of the sting be left behind, and if so, it should be carefully pulled out with a pair of tweezers. If, however, a bit of the sting be broken off and left in, and it be incautiously rubbed, this increases the mischief; the sides of the wound should then be gently squeezed, if possible to push it out. When the wound is thus cleared it is best to anoint the part with sweet oil, which generally relieves the pain ; the swelling, however, does not subside for some hours. If the wound become angry and tender, it is best to apply a poultice.*

In eating summer fruits people should be cautious that no lurking wasp be swallowed, as instances of death are not wanting in consequence of the insect stinging the gullet.

Adder or Viper Bite.

This is a very serious accident, and in some parts of the country not unfrequent in very hot weather if the reptile be suddenly disturbed by cutting furze or brushwood, or by being trod on, or if an attempt be made to catch hold of it whilst endeavouring to make its escape, which it

---

* A French practitioner recommends, in case of these stings, to rub the part with garlic.

ıs rather inclined to do than to attack. The
following relation of a case will perhaps be the
best mode of describing the symptoms and treatment of this injury :—

Many years ago I was returning, by steamboat, from a short sojourn in the lovely Weald
of Kent, and amusing myself, as the evening
drew on, with watching the varied groups of
weary pleasure-hunters by whom I was surrounded, when a young man pushed through the
crowd, civilly pulled off his hat, hoped I would
excuse him, and begged I would go down into the
cabin, and see a friend who was very ill, having
been bitten a few hours before by an adder,
which one of his companions had picked up in a
field near Gravesend. They had killed the ugly
animal and rubbed some of its fat upon the
wound, but without much benefit ; and when they
had reached the town, their friend was so ill that
they took him to a doctor's and got some stuff
for him, which was equally useful as the fat. He
became very faint and vomited, and therefore,
as the best they could do, had got into the first
boat and were bringing him to town, but as he
continued getting worse they became frightened,
and hence the request that I would see him.

Into the cabin I accordingly went, and found
the man " pale as his shirt ; his knees knocking ;"
with anxious countenance and " lack-lustre eye,"
and voice scarce to be heard above a whisper, in

which he told of agonizing pain extending up his
arm. He had been bitten on the back of his
thumb, and the wound was such as a moderate-
sized pin would make. The back of the hand had
swollen as high as the wrist, was pale as marble,
but not very tender, and the swelling was evidently spreading up the arm.

A steamboat was not a very likely place to
get 'pothecary's stuff, but it readily furnished
what was needful, brandy and oil ; a glass or
two of the former was given to rouse the patient
from his fainting state, and the hand and arm
were gently and continually anointed, by the
finger, with the latter. By these means he was a
little roused and freed from pain, but the swelling continued, and in the course of a few hours
had reached the shoulder. During the night
the oiling was continued, and more brandy occasionally given till the faintness went off. Next
morning he was tolerably free from pain, but
the whole limb was hide-bound, pale, and doughy,
and the wound scarcely to be seen. A poultice
was, however, applied to it, and the oiling continued. Next day there was a little discharge
from the wound, the swelling of the arm was sub-
siding, in two or three days it entirely disappeared, and he was well.

Such is the history of an adder or viper bite,
and the cases I have seen have all had nearly
the same character, though sometimes the effects

are more severe, and violent griping of the bowels comes on soon after the accident.

The sequel of the story was rather a curious one, as I found on visiting him next day at the hospital. He and his friends were a party of pickpockets: his accident spoilt their sport, and the biter was bitten.

Restoration of the patient's quickly exhausted powers by brandy, Cologne water, sal-volatile and water, or any other stimulant at hand, is first needed; and then continual anointing with sweet oil generally completes the cure. It is a vulgar notion that the adder carries its own antidote in the fat contained in its belly. With this idea it is the common practice to kill the animal, and smear his grease upon the part, and perhaps, if no other remedy can at once be got, the application is not a bad one.

ASTLEY COOPER and EVERARD HOME both recommended tying a string tightly round the limb above the bitten part, so as to prevent the poison running up among the loose parts immediately beneath the skin. I have no experience of this mode of treatment; it might be done, perhaps effectually, if a finger or toe be wounded, but it could hardly be managed if the leg or arm were the injured part.

I would not advise sucking the wound with the mouth to get the poison out, for if by chance the sucker should have a cracked lip or any

slight sore in his mouth, he would assuredly poison
the wound, though less severely.

It may not be out of place to notice that in
this country there is only one poisonous of the
snake kind, the *Adder* or *Viper*, which varies from
reddish black to nearly black, and is of less size
than the common harmless snake, which is readily distinguished by a bright yellow collar just
behind the head, and though it hisses, does not
bite. Adders are fond of warmth. SHAKE-
SPEARE speaks of the "bright day that brings
forth the adder;" and in the hot months of
summer it is active, and its poison more virulent;
but in cold weather it is torpid or sluggish, and
its bite less severe. It should, however, even
then be treated with caution, as the warmth of
the hand will quickly excite it, to the inconvenience of the incautious holder. This happened to
ASTLEY COOPER, who, having frozen an adder
to show the effects of cold upon it, and continuing
his lecture, forgetful that he held the venomous
reptile in his hand, was bitten by it, on its revival, through the finger.

Adder-bites are rarely fatal, but instances are
not wanting in which persons have died. One is
mentioned of a woman who died thirty-seven
hours after, and another, in which erysipelas and
sloughing of the limb having followed, a young
man died some weeks afterwards.

The adder inflicts its bite by means of a pair of

very sharp curved teeth or fangs ($a\,a$) attached to
the upper jaw, and when at
rest lie along each side of the
mouth with their points backwards, and contained in a
sheath of skin (b), in which
they are concealed ; but when
the animal is disposed for at-
tack it has the power of bringing them down vertically to the mouth, and then striking them
forcibly into the animal it attacks.

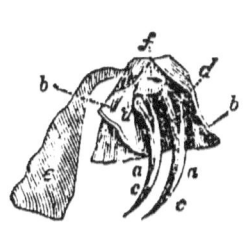

Correctly speaking, there are in each sheath
two fangs, as seen in the cut, and sometimes
more, so that in case of accident the animal has
one in reserve, but it only uses one of them ; and
what is still more remarkable, as far as I have
seen, it does not use the fangs of both, but only
of one side. The fang is hollow, and has two
small openings, one ($d\,d$) a little below its root,
through which the poison is thrown from a
gland (e) placed at the back of the jaw, and
communicating with it by a delicate soft tube
(f); and the other ($c\,c$) just above its point,
by which it flows into the wound.

In all the poisonous snakes which are met with
abroad the poison is conveyed in much the same
way through a hollow tooth; but the poison-
tooth or fang is not always situated in the front
of the mouth, and hence, from this having been
unknown, many snakes have been considered
harmless which are actually very venomous.

Rattle-snake Bites

Produce much the same symptoms as those by
adders. According to BARTON,* though rarely,
they are sometimes fatal, either within a few hours,
or after some days in consequence of the sloughing, which the constitution has not the power to
withstand. He observes, "In those cases where
the poison is applied near the orifice of an absorbing vessel, we have reason to suppose that it will
be conveyed into the mass of blood with great ce-
lerity. . . . But, unfortunately, cases sometimes
occur in which this active matter is thrown immediately into a vein or artery. When this happens,
the effects of the poison will be more readily pro-
pagated to the remotest part of the system." The
following are the symptoms which occur after
this accident as stated by BARTON :—" When the
poison of the rattle-snake has actually been introduced into the general mass of blood, it begins to
exert its most alarming and characteristic effects.
A considerable degree of nausea (disposition to
vomit) is a very early symptom. We now discover an evident alteration in the pulse ; it becomes full, strong, and greatly agitated. The
whole body begins to swell; the eyes become
so entirely suffused that it is difficult to discover the smallest portion of the adnata (the
general covering of the eyeball and eyelids) that
is not painted with blood. In many instances

---

* American Philosophical Transactions, vol. iii. 1793.

there is an hæmorrhage of blood from the eyes, and likewise from the nose and ears; and so great is the change induced in the mass of blood, that large quantities of it are sometimes thrown out on the surface of the body in the form of sweat; the teeth vacillate in their sockets, whilst the pain and groans of the unhappy sufferer too plainly inform us that the extinction of life is at hand. In this stage of its action, and even before it has induced the most alarming symptoms which I have mentioned, the powers of medicine can do little to check the rapid and violent progress of this poison."

The treatment adopted is thus described by the same writer :—" In general the first thing that was attended to after a person had been bitten by the rattle-snake was to throw a tight ligature above the part into which the poison had been introduced ; at least this was the practice whenever the situation of the wounded part admitted of such an application. The part was next scarified, and a mixture of salt and gunpowder, sometimes either of these articles separately, was laid upon the part. Over the whole was applied a piece of the bark of the white walnut. At the same time some one, frequently more than one, of the vegetables which were mentioned to me,* were given internally,

---

* A vast number, not worth mentioning, however, as BARTON does not rely on any of them.

either in decoction or infusion, along with a large quantity of milk."

Hooded Snake's (*Cobra de Capello*) bite.

The symptoms as described by PATRICK RUS-SELL,* differ from those caused by the rattle-snake. In one instance, a woman who was bitten, "after ten hours had lost her senses of seeing and feeling, and deglutition (swallowing) was so much impeded, that hardly anything could be got to pass into the stomach. No other parts were visibly affected by spasms; but a torpor and listlessness pervaded the whole system, and from the moment of the bite had continually increased." In another case, immediately after the bite, the man was "sensible of pain. In half an hour the pain had extended up to the knee, and ten minutes after up to the top of the thigh, and much more severe. He then complained of severe pain in the belly, which was tense and much swollen. A sense of tightness spread towards the chest, and respiration (breathing) became very laborious. Soon after deglutition became impeded, and the stricture in the œsophagus (the gullet) increased so much that nothing could be forced down his throat; he foamed at the mouth; his eyes stood staring and fixed; his pulse and respiration became hardly perceptible, and, in short, every vital motion seemed at a stand."

---

* An Account of Indian Serpents, &c. 1796.

The treatment of such cases was with the celebrated Tanjore pill, which consisted of " white arsenic, roots of vellinavi, kernels of nervalam, pepper, quicksilver, of each an equal part. The quicksilver is to be rubbed with the juice of the wild cotton (*Asclepias gigantea*) till the globules become invisible. The arsenic being first levigated, and the other ingredients reduced to a powder, are then added, and the whole is beaten up together with the juice of the wild cotton to a consistence fit to be divided into pills. If a person is bit by a cobra de capello, mix one of the pills with a little warm water and give it to the patient. After waiting a quarter of an hour, should the symptoms of infection increase, give two pills more ; should these not sufficiently counteract the poison, another pill must be given an hour after. The wound should be dilated, and the warm liver of a fowl applied to the part. This is generally found sufficient. The patient ought to keep a regimen for six days, eating only congee (rice water) and rice, or milk and rice. He should abstain from salt, and his drink may be warm water. Sleep is to be prevented for the first twenty-four hours. The pills gene-rally occasion a nausea and purging, but seldom in a violent degree."

Bite of a mad Dog.

The consequences to a person bitten by a mad

dog, cat, or other animal, if not attended to, are so horrible and so certainly fatal—although the length of time at which hydrophobia appears may vary from a few days to eighteen or twenty months, though more commonly it makes itself known in the course of six or twelve weeks—that it is of the utmost importance no time should be lost in using the most probable means to ensure the person's safety, though occasionally even these are unsuccessful.

Treatment.

If a medical man can be found, *have the bitten part cut out, and never be persuaded to have less done;* he may do what more he thinks proper. And if the animal's tooth have entered to any depth, *let a probe or a bit of stick be passed down to the bottom of the wound, and insist on the operator cutting around the wound, and so deeply as to bring out the probe or stick covered with the part cut out as with the finger of a glove.* This will ensure safety, but it is usual afterwards to apply caustic in the new wound which the surgeon has made, and it is well to do this. If the lip or cheek be bitten through, have the piece cut completely out into the mouth. No hesitation must be had about spoiling the beauty; the only consideration is to save the person from a certain and most dreadful death.

If the skin be only grazed by the animal's

tooth, have it cut out to at least a quarter of an inch distance around the slightest appearance of the graze, and then have caustic applied.

If a doctor cannot be at once found, never fail, even though the wound have healed, as soon as he can be met with, to have the wounded part, or even the scar, completely cut out, as some think that the poison remains for a less or a greater time in the wound without affecting the constitution, and therefore that the scar ought to be removed at any time before the symptoms come on. Therefore never neglect even this.

I do not think I am wrong in advising, where the finger has been bitten and there is no possibility of medical assistance, to chop it off, which any one can do with a mallet and chisel. The danger is so urgent, the consequences so dreadful, that any mutilation is warrantable, and should be submitted to in hope of escaping this frightful disease.

Caustic alone should never be relied on, if it be possible to remove the injured parts with the knife, and there are very few parts where the knife cannot be employed.

Never trust any of the remedies which have been said to be capable of preventing hydrophobia short of the knife and caustic. As yet there is *no authentic account of any remedy* except the knife and caustic having been effectual for dog-madness.

K

It is the almost general practice, if any one
have been bitten by a strange dog, to raise the
cry "mad dog," in consequence of which the
poor animal is summarily dispatched or hunted
to death.   This, for the sake of the person bitten,
should always, if possible, be prevented.   Because frequently the dog is not mad, and therefore ascertaining this fact is of the utmost importance to relieve the patient's mind.   Whilst,
on the other hand, if the dog should prove to be
mad, means may be taken to prevent further
mischief from other animals among which the
dog may have been accustomed to be, or among
which it may have gone and bitten them. The dog
should therefore be properly secured, and carefully
watched for some weeks or even months.   If the
dog have been destroyed, it should be carefully
examined by some person fully conversant with
the peculiar appearances which the throat and
stomach present.   But if there be any doubt on
the point, I think it is advisable to test the saliva
or spittle, in which the poison exists, by inocu-
lating some other dog, and then putting him in
some safe place to watch the results.

I have not thought it necessary to give any
detail of the frightful symptoms which render
hydrophobia in the human subject as horrible as
it is a certainly fatal disease.   No object would
have been attained by such dreadful recital.
This, however, does not apply to some account

of the disease in the dog; of which, on the contrary, it is of great importance that people have a good general knowledge, so that they should not be needlessly terrified, as they too frequently are, nor be endangered from carelessness or ignorance of the signs by which madness in dogs shows itself. I have, therefore, very largely extracted from YOUATT's excellent work, 'The Dog,' his account of this fearful disease as it presents itself in that animal.*

" The early symptoms of rabies in the dog are occasionally very severe. In the greater number of cases these are sullenness, fidgetiness, and continual shifting of posture. Where I have had opportunity, I have generally found these circumstances in regular succession. For several consecutive hours, perhaps, he retreats to his basket or his bed. He shows no disposition to bite, and he answers the call upon him laggardly. He is curled up, and his face is buried between his paws and his breast. At length he begins to be fidgety. He searches out new resting-places; but he very soon changes them for others. He takes again to his own bed, but he is continually shifting his posture. He begins to gaze strangely about him as he lies on his bed. His

---

* I have to offer my best thanks to Mr. CHARLES KNIGHT for his kind permission to make use of such part of ' The Dog' as I have required for the present purpose.

K 2

countenance is clouded and suspicious. He comes to one and another of the family and he fixes on them a steadfast gaze, as if he would read their very thoughts. . . . . . . A peculiar delirium is an early symptom, and one that will never deceive. . . . . . I have again and again seen the rabid dog start up, after a momentary quietude, with unmingled ferocity depicted on his countenance, and plunge with a savage howl to the end of his chain. At other times he would stop and watch the nails in the partition of the stable in which he was confined, and, fancying them to move, he would dart at them, and occasionally sadly bruise and injure himself from being no longer able to measure the distance of the object." But, " whether he is watching the motes that are floating in the air, or the insects that are annoying him on the walls, or the foes that he fancies are threatening him on every side— one word recalls him in a moment. Dispersed by the magic influence of his master's voice, every object of terror disappears, and he crawls towards him with the same peculiar expression of attachment that used to characterise him. Then comes a moment's pause—a moment of actual vacuity—the eye slowly closes, the head drops, and he seems as if his fore feet were giving way and he would fall; but he springs up again, every object of terror once more surrounds him— he gazes wildly around—he snaps—he barks, and he rushes to the extent of his chain, prepared to meet his imaginary foe.

"The expression of the countenance of the dog undergoes a considerable change, principally dependent on the previous disposition of the animal. If he was naturally of an affectionate disposition, there will be an anxious, inquiring countenance, eloquent beyond the power of resisting its influence. It is made up of strange suppositions as to the nature of the depression of mind under which he labours, mingled with some passing doubts, and they are but passing, as to the concern which the master has in the affair; but, most of all, there is an affectionate and confiding appeal for relief. At the same time we observe some strange fancy, evidently passing through his mind, unalloyed, however, by the slightest portion of ferocity.

"In the countenance of the naturally savage brute or him that has been trained to be savage there is indeed a fearful change; sometimes the conjunctiva is highly injected; at other times it is scarcely affected, but the eyes have an unusually bright and dazzling appearance. They are like two balls of fire.

"A very early symptom of rabies in the dog is an extreme degree of restlessness. Frequently he is almost invariably wandering about, shifting from corner to corner, or continually rising up and lying down, changing his posture in every possible way, disposing of his bed with his paws, shaking it with his mouth, bringing it to a heap, on which he carefully lays his chest, or rather the pit of his stomach, and then rising up and bundling every portion of it out of the kennel. If he is put into a closed basket he will not be still for an instant, but turn round and round without ceasing. If he is at liberty he will seem to imagine that something is lost, and he will eagerly search round the room, and particularly every corner of it, with strange violence and indecision.

"The ear is oftener than any other part bitten by the rabid dog, and when a wound in the ear, inflicted by a rabid dog, begins to become painful, the agony appears to be of the intensest kind. The dog rubs his ear against every projecting body, he scratches it might and main, and tumbles over and over while he is thus employed. Is this dreadful itching a thing of yesterday, or has the dog been subject to canker, increasing for a considerable period? Canker, both internal and external, is a disease of slow growth, and must have been long neglected before it will torment the dog in the manner I have described. The question as to the length of time that an animal has thus suffered will usually be a sufficient guide. The mode in which he expresses his torture will serve as another direction. He will often scratch violently enough

when he has canker, but he will not roll over
and over like a football except he is rabid.

"In the early stage of rabies the attachment
of the dog towards his owner seems to be rapidly
increased, and the expression of that feeling.
He is employed, almost without ceasing, licking
the hands, or face, or any part he can get at.
Females, and men too, are occasionally apt to
permit the dog, when in health, to indulge this
filthy and very dangerous habit with regard to
them.

"A depraved appetite is a frequent attendant
on rabies in the dog. He refuses his usual food:
he frequently turns from it with an evident expression of disgust; at other times he seizes it
with greater or less avidity, and then drops it,
sometimes from disgust, at other times because
he is unable to complete the mastication (chew-
ing) of it. This palsy of the organs of mastication and dropping of the food, after it has been
partly chewed, is a symptom on which implicit
confidence may be placed. Some dogs vomit
once or twice in the early period of the disease:
when this happens they never return to the
natural food of the dog, but are eager for everything that is filthy and horrible. The natural
appetite generally fails entirely, and to it succeeds a strangely depraved one. The dog
usually occupies himself with gathering every
little bit of thread, and it is curious to observe

with what eagerness and method he sets to work, and how completely he effects his object. He then attacks every kind of dirt and filth, horse-dung, his own dung, and human excrement. Some breeds of spaniels are very filthy feeders without its being connected with disease, but the rabid dog eagerly selects the excrement of the horse and his own. Some considerable care, however, must be exercised here," because at particular times, and especially whilst cutting his teeth, " a young dog will be observed gathering up hard substances, and if he should chance to die, a not inconsiderable collection of them is sometimes found in the stomach. They are, however, of a peculiar character ; they consist of small pieces of bone, stick, and coal. The contents of the stomach of the rabid dog are often, or generally, of a most filthy description. Some hair or straw is usually found, but the greater part is composed of horse-dung or of his own dung, and it may be received as a certainty that if he is found deliberately devouring it, he is rabid.

" The dog at particular times may, and does, diligently search the urining-places; he may even, at those periods, be seen to lick the spot which another has just wetted : but if a peculiar eagerness accompanies this strange employment ; if, in the parlour, which is rarely disgraced by this evacuation, every corner is perseveringly examined and licked with unwearied and unceasing industry,

that dog cannot be too carefully watched : there
is great danger about him ; he may, without any
other symptom, be pronounced to be decidedly
rabid. I never knew a single mistake about this.

" Much has been said of the profuse discharge
of saliva (spittle) from the mouth of the rabid
dog. . . . But it never equals the increased
discharge that accompanies epilepsy, or nausea
(a disposition to be sick). The frothy spume (in
rabies) is not for a moment to be compared with
that which is evident enough in both these affec-
tions. It is a symptom of short duration, and
seldom lasts longer than twelve hours. The
stories that are told of the mad dog covered with
froth, are altogether fabulous. The dog recover-
ing from, or attacked by a fit, may be seen in this
state; but not the rabid dog. Fits are often
mistaken for rabies, and hence the delusion.

" The increased secretion of saliva soon passes
away. It lessens in quantity ; it becomes thicker,
viscid, adhesive, and glutinous. It clings to the
corners of the mouth, and probably more annoy-
ingly so to the membrane of the throat. . . . .
The dog furiously attempts to detach it with his
paws. It is an early symptom in the dog, and
it can scarcely be mistaken in him. When he
is fighting with his paws at the corners of his
mouth, let no one suppose that a bone is sticking
between the poor fellow's teeth ; nor should any
useless and dangerous efforts be made to relieve

him.   If all this uneasiness arose from a bone in
the mouth, the mouth would continue perma-
nently open, instead of closing when the animal
for a moment discontinues his efforts.   If after a
while he loses his balance, and tumbles over,
there can be no longer any mistake.   It is the
saliva becoming more and more glutinous, irritating the throat, and threatening suffocation.

"To this naturally and rapidly succeeds an
insatiable thirst.   The dog that still has full
power over the muscles of his jaws continues to
lap.   He knows not when to cease; while the
poor fellow, labouring under the dumb madness,
presently to be described, and whose jaw and
tongue are paralyzed, plunges his muzzle into
the water-dish to his very eyes, in order that he
may get one drop of water into the back part of
his mouth to moisten and to cool his dry and
parched throat.   Hence, instead of this disease
being always characterised by the dread of water
in the dog, it is marked by a thirst often perfectly
unquenchable.

"In almost every case in which the dog utters
any sound during the disease, there is a manifest
change of voice.   In the dog labouring under
ferocious madness, it is perfectly characteristic.
There is no other sound that it resembles.   The
animal is generally standing, or occasionally
sitting, when the singular noise is heard.   The
muzzle is always elevated.   The commencement

is that of a perfect bark, ending abruptly and
very singularly, in a howl a fifth, sixth, or eighth
higher than at the commencement.   Dogs are
often enough heard howling ; but in this case it
is the perfect bark, and the perfect howl rapidly
succeeding to the bark.

·   " There is another partial change of voice, to
which the ear of the practitioner will by degrees
become habituated, and which will indicate a
change in the state of the animal quite as dangerous as the dismal howl ; I mean when there
is a hoarse inward bark with a slight but characteristic elevation of the tone.   In other cases,
after two or three distinct barks will come the
peculiar one mingled with the howl.   .Both of
them will terminate fatally, and in both of them
the rabid howl cannot possibly be mistaken.

" There is a singular brightness in the eye
of the rabid dog, but it does not last more than
two or three days.   It then becomes dull and
wasted ; a cloudiness steals over the conjunctiva,
which changes to a yellow tinge, and then to a
dark green, indicative of ulceration deeply seated
within the eye.   In eight and forty hours from
the first clouding of the eye, it becomes one dis-
organized mass.

" Absence of pain in the bitten part is an
almost invariable accompaniment of rabies.   I
have known a dog set to work, and gnaw and tear
the flesh completely away from his legs and feet.

" In the great majority of cases of furious madness, and in almost every case of dumb madness, there is a staggering gait, not indica‑ tive of general weakness, but referable to the hind quarters alone, and indicating an affection of the lumbar motor nerve. In a few cases it approaches more to a general paralytic affection.

" It is not every dog that in the most aggra‑ vated state of the disease shows a disposition to bite. . . . On the other hand, there are rabid dogs whose ferocity knows no bounds. It they are threatened with a stick, they fly at, and seize it, and furiously shake it. They are in‑ cessantly employed in darting to the end of their chain, and attempting to crush it with their teeth, and tearing to pieces their kennel, or the woodwork that is within their reach. They are regardless of pain. The canine teeth, the incisor teeth, are torn away; yet, unwearied and insensible to suffering, they continue their efforts to escape.

" If by chance a dog in this state effects his escape, he wanders over the country, bent on destruction. He attacks both the quadruped and the biped. He seeks the village street, or the more crowded one of the town, and he suffers no dog to escape him. The horse is his frequent prey, and the human being is not always safe from his attack. . . . At length the rabid dog becomes completely exhausted, and slowly reels along the road with his tail depressed, seemingly half unconscious of surrounding objects. His open mouth, and protruded and blackened tongue, and rolling gait, sufficiently characterise him. He creeps into some sheltered place, and then he sleeps twelve hours or more. It is dangerous to disturb his slumbers, for his desire to do mischief immediately returns, and the slightest touch or attempt to caress him is repaid by a fatal wound. This should be a caution never to meddle with a sleeping dog in a wayside house, and, indeed, never to disturb him anywhere.

" In an early period of the disease, in some dogs, and in others when the strength of the animal is nearly worn away, a peculiar paralysis of the muscles of the tongue and jaws is seen. The mouth is partially open, and the tongue protruding. In some cases the dog is able to close his mouth by a sudden and violent effort, and is as ferocious and as dangerous as one, the muscles of whose face are unaffected. At other times the palsy is complete, and the animal is unable to close his mouth or retract his tongue. (I presume this is the form of disease which YOUATT calls " dumb madness," but he nowhere positively mentions it.—J. F. S.)

" In the dog, I have never seen a case in which plain and palpable rabies occurred in less than fourteen days after the bite. The average time I should calculate at five or six weeks. In three months I should consider the animal as

tolerably safe.   I am, however, relating my own
experience, and have known but two instances in
which the period much exceeded three months.
In one of these five months elapsed, and the
other did not become affected until after the expiration of the seventh month.

" The duration of the disease is different in
different animals.   In man it has run its course
in twenty-four hours, and rarely exceeds seventy-
two.   In the horse, from three to four days; in
the sheep and ox, from five to seven; and in the
dog, from four to six."

TORN OR CUT ACHILLES' TENDON.

THE large thick tendon thus called, which con-
nects the heel with the great muscles forming the
calf of the leg, and which are very main parties
in keeping the lower limbs erect when we stand,
and in throwing the body forward as we walk, is
liable to be torn or cut.

This tendon is liable *to be torn* in making a
false step whilst walking, or in coming down
stairs. It is also said to be torn whilst people
are exerting themselves energetically in dancing.

The tendon tears without warning, and the
person drops to the ground as if shot. Should the
accident happen whilst walking, he has the sensation as if some mischievous fellow had struck
him with a stone, but on looking about he finds
no one near him. When he gets up he finds
himself utterly unable to keep that leg erect if he
make the least attempt to rest his weight on it,
and is therefore compelled to hop on the other.

Treatment.

This consists in putting the person to bed,
and laying his leg on the outside, with his knee
much bent, and the toes much pointed, by
which position the torn ends of the tendon are
brought as nearly together as possible. This

posture must be preserved for about a fortnight, to give time for the production of the new substance by which the tendon is to be repaired; and as it can scarcely be constantly kept up without restraint it is better to put a piece of thin board, about three fingers wide, and extending from below the knee-cap beyond the toes, upon the front of the leg, taking care to have the board well padded with three or four thicknesses of rug or thick flannel, so that it may not rub.   It must

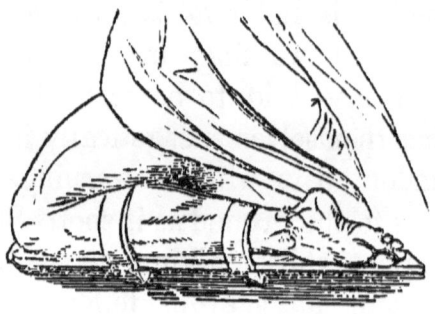

be confined above by a few turns of a short roller around it and the upper part of the calf; and below, around it and the foot, so that the pointing of the toes is thus rendered continual.   No bandage must be put on at the part where the tendon has been torn, and which is easily found before the foot is extended, by the gap into which the finger drops in passing it from the heel up the leg towards the calf.   .

After a fortnight, or it may be a little longer, has passed, on feeling for the gap, its place is found filled with a firm substance.   The person may then get up, and begin to move about a little ; his shoe, or rather a laced half-boot, for the shoe he will not be able to keep on, being provided with a high cork heel, which should keep the toes nearly as much pointed as whilst he was in bed with the board on the front of the leg. After a week, a thin slice of the cork heel may be taken off, and subsequently it may be lowered more and more, till at last the heel can bear upon the ground as usual.

Another mode is sometimes adopted abroad, of bending the leg upon the thigh, and tying it up with the toes upwards. This is doing precisely the same as that just directed, but it is awkward and inconvenient.

Sometimes the Achilles' tendon is *cut through* by the adze slipping whilst a carpenter is smoothing a floor, or it may be divided with a scythe, when one man is mowing too closely to another. The person thus maimed is exactly in the same condition as a beast which is houghed.

This is a much worse accident than the torn tendon, as in placing the limb in the same posture, which is also requisite to bring the cut ends of the tendon together, the loose skin drops into the wound, and gets between its edges, so as to interfere with their union. The edges of the skin must be kept together by two or three stitches of silk, and instead of passing the thread through from the lower to the upper part of the divided skin,

L

whilst its edges are simply brought together, it is best to nip up both edges of the skin so as to make their under surface touch, and then pass the needle and thread upwards through both together, about two tenths of an inch from the edge, and then at a quarter of an inch distance to pass it again downwards in like way,—in fact, to make one "running stitch," as it is called by women. Two or more such stitches, according to the size of the wound, must be put in, and should be supported by long narrow strips of sticking plaster laid between them lengthwise on the leg. About the third or fourth day the stitches must be taken out, if the holes through which the needles have passed be wet with matter, or before this time if they be red and angry, and the threads seem to be cutting the skin, as they are then not merely useless from ceasing to give support, but also become actual sources of irritation. After they are removed, the straps of plaster must be used to keep the wound together.

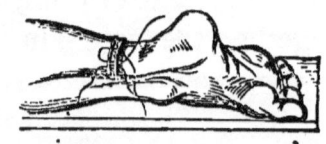

WHAT IS TO BE DONE IN CASES OF SUDDEN BLEEDING FROM VARIOUS CAUSES.

BLEEDING may take place suddenly from over-loading of the vessels with blood, or in consequence of wounds; and both may be fatal, if unchecked, although those from the latter cause are generally more quickly so than the former.

Bleeding from overloaded Vessels.

One of this kind of bleedings has already been noticed (p. 87)—that in which a swollen vein in the leg gives way.

Bleeding *from the nose* is very frequent in young people. Generally this is checked by the person sitting upright, bathing the nose externally with cold water, or vinegar and water, and sniffing it up the nostrils. If, however, it continue, a moderate pinch of powdered alum may be put into a couple of tablespoonfuls of water, and thrown up with a squirt; or a plug of lint dipped in this wash may be passed into the bleeding nostril, for generally it is only one side which does bleed; but care should be taken to fasten a strong thread securely round it, lest it be pushed in or slip so far back into the nostril

L 2

that it cannot be got out without much difficulty. Where there is frequent disposition to
bleeding from the nostrils, it is necessary to
prevent costiveness, and to take some saline
purge continually, so as to keep the bowels
rather relaxed.

Blood may be coughed up *from the lungs*,
or vomited *from the stomach*, both which are
very serious matters, and require immediate
attention.

When blood is *coughed up*, if in small quantity, it is shown to come from the lungs by its
frothiness; yet if in large quantity it is not frothy,
but pure bright red blood. The person should
be at once put to bed, in a cold room, and kept
cool. If faint, no attempt to recover him by
giving wine, brandy, or other stimulants, should
be made, which will only keep up the bleeding
and increase the patient's danger, which is lessened by the faintness. If no medical man can
be quickly obtained, but there be any one near
who can either cup or bleed, the patient should
be cupped on the chest, or bled from the arm,
forthwith, to the amount of a pint, or a pint and
a half, according to his strength, which it is
the object to reduce, so as to lessen the power
of the heart, and thereby diminish or check the
continuance of the pouring out of the blood into
the lungs. Three grains of calomel may be
given at once, but nothing taken except cold

or iced drinks, till the medical man arrive and pursue such further treatment as the case requires.

If there be no doubt of the blood coming from the lungs it would be very proper to give two or three grains of sugar of lead with a quarter of a grain of opium made into a pill with breadcrumb every four or six hours.

When blood is *vomited from the stomach*, it is known by its dark colour, and by being mixed with the contents of the stomach ; and it is usually preceded by weight, pain, and uneasiness in the region of the stomach.

In this case, also, the person must be put to bed, kept cool and quiet, and, as the cause of the bleeding varies, often, though not always, bleeding from the arm is required or from the region of the stomach by cupping or leeches. Cold liquids, or ice, in small quantities only, may be given till the medical attendant comes to take charge of the case.

Bleeding from Wounds.

Bleeding from a wound may in general be for a time, or even completely, stopped, if the wounded part be on a bone, as for instance on the skull, or on parts of the face, where it can be pressed firmly against the bone by the finger, or by a bit of cork, or a hard pad bound tightly on with a roller. But if this do not succeed, each edge of the wound may be lifted up, carefully examined, and if any small jet of blood be seen, it may be presumed that some little artery is wounded. The point of a tenaculum (A) should then be dipped in as near as possible to it, and the spouting mouth (B) drawn up sufficiently to pass a strong thread or silk round it, below the tenaculum, one end of the silk should then be passed through the other, and 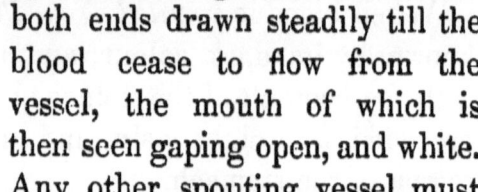 both ends drawn steadily till the blood cease to flow from the vessel, the mouth of which is then seen gaping open, and white. Any other spouting vessel must be hooked up and tied in the same way. After which, if the bleeding cease, the wound may be brought together with plaster.

When, however, wounds happen in the limbs, and are followed by much and continued bleeding, it cannot usually be stopped by pressure on or near the wound, but requires the whole current of blood to be prevented passing through the limb. This is very easily done, and an unprofessional person can be taught with little difficulty to do it.

If the bleeding be from a wound in the arm, more especially if near the armpit, in which case nothing more can for the moment be done, a by-stander should press his thumb firmly into the neck, behind the middle of the collar-bone, which will stop the flow of blood through the great artery of the arm as it is first coming out of the chest. As, however, the pressure thus made soon tires the thumb, the handle of a door-key,

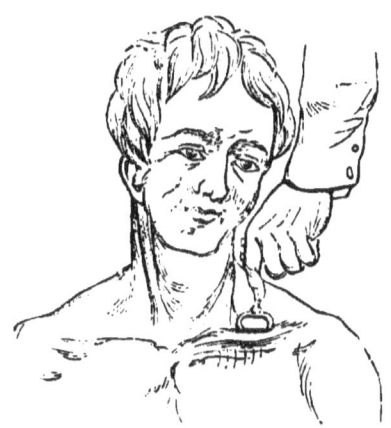

wrapped in three or four folds of linen, may be pressed behind the middle of the collar-bone, and held without fatigue for almost any length of time, till proper assistance can be obtained.

If there be very severe bleeding from a wound in the leg or thigh, especially if high up in the latter, the great artery which supplies blood to the limb may be pressed so as to prevent the flow of blood through it by pressing with the thumb immediately below the crease of the groin. This pressure is made with less difficulty than in pressing behind the collar-bone, because the patient lying on his back, the pressure is made directly upon the groin, at right angle with the body. The door key may be here also used, but the thumb is the most convenient.

When, however, the bleeding wound is any-where below the middle of the upper arm, or below the middle of the thigh, a temporary con-trivance may be used which will command the bleeding.   It consists merely of a stout pocket-handkerchief and a piece of tough stick, which, from its mode of employment, is called a stick-tourniquet.

The handkerchief is to be passed once or twice round the limb, some distance, if possible, above the wound, and tied tightly and firmly. The stick is then pushed beneath the circular bandage thus formed, between it and the skin, and twisted so that it screws the handkerchief tight till the blood cease to flow.   The screwing should only be continued till the bleeding stop; for if the bandage and stick be strong, and the twisting be continued, the soft parts beneath may be severely and unnecessarily bruised.

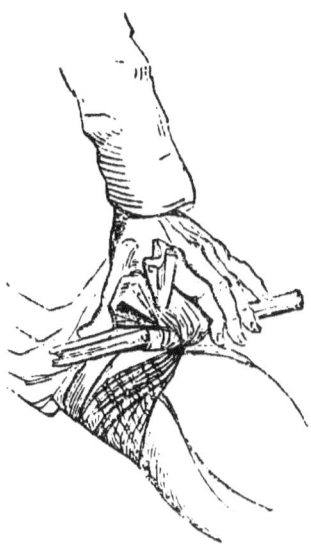

Surgeons have a brass tourniquet with a bandage and a pad, the action of the pad being
to press specially upon the artery ; but as the
situation of the vessel requires anatomical
knowledge, it is not useful to an unpractised
person, and the stick tourniquet, applied as directed, and screwed tight, answers the purpose
exceedingly well, till proper assistance can be
obtained.

SCALDS AND BURNS.

SCALDS and burns are among the most common accidents; and the mischief or even danger which results from them is greatly increased by the want of immediate attention. It is therefore of great importance that every one should know how to make use, at once, of one or other of the means which most houses are provided with, or which can be obtained in a few minutes from a neighbour.

Scalds from hot water, unless very extensive, are generally least severe, because the scarf-skin only is raised, as in a common blister. Scalds from boiling oil or varnish are more severe, because, as they stick to the skin, the heat is longer applied, and the true skin is therefore more likely to be destroyed, either without or with a blister being formed.

Burns are generally most serious, because fire being applied to the skin, if its application be continued for a few minutes, it roasts or chars the skin and soft parts beneath it, and produces a slough or mortification of the part, which, however large it may be, must come away before any progress towards repair of the damage can be made : and not unfrequently, if the mortified part

be large, the person has not strength to contena
with this process, and dies at last, although he
may have lingered for weeks.

Scalds and burns are less or more dangerous,
also, as regards the part injured. Children occasionally will suck the spout of a kettle boiling
or nearly boiling. This is most dangerous, especially if the steam be drawn into the windpipe, the lining of which almost immediately
swells, and the child is suffocated in a very few
hours. Upon the trunk, and especially upon the
chest, and if in children, scalds and burns are
almost invariably fatal. They are also more
dangerous on the lower than on the upper
limbs and face. The consequences of severe
scalds and burns, however judiciously treated
and however well they may, for a time, seem to
go on, are often very lamentable. Not only
are very disfiguring scars produced, but these
scars have for a long while a disposition to contract powerfully, and, dragging the neighbouring
parts together, to fix them. Thus, if the front of
the arm be scalded or burnt so that a scar is produced about the bend of the elbow, the forearm
and hand will be often drawn up to a right angle,
from which they cannot be straightened, and a
large web of skin extends from one to the other.
So, if the front of the neck be scalded, the chin
will be drawn down to the chest, the mouth kept
wide open, and the lower lip turned out, in con-

sequence of which there is continual drivelling of
the spittle, and not unfrequently the neck becomes
sore.   The medical attendant is often blamed
for these sad occurrences, but most unjustly, for
it is out of his power to prevent them, do whatever he may.

It hardly need be observed when a person is
scalded, that the part scalded must immediately
be got away from the scalding liquor, whatever it
may be ; but more commonly this is not required,
as more scalds happen from hot or boiling fluids
falling upon, than from a person falling into them.

But as regards burns, the first thing to be
done is to put the blazing fire out, if the accident
have occurred by the clothes catching fire.
Much of the danger in these cases depends on
the terrified person running about, and thus actually fanning the flame, and increasing the fire
in which he or she (the latter especially, as is
more commonly the case) is enveloped.   If the
person have sufficient presence of mind, he should
throw himself down on the floor and roll over
and over till the flames be put out ; and if there
be a rug or loose carpet in the room, he should
roll himself up in it as quickly and completely as
possible.   By thus doing the fire will generally
be put out, or its progress checked till assistance, with buckets of water, can be obtained.

The sufferer having been got away from the
scalding water, or the fire having been quenched,

he should be put to bed as quickly as possible, and the clothes removed from the scalded or burnt part. Saving the clothes should not for a moment be thought about ; everything, excepting the body-linen or flannel, should be cut away, so that it may drop piecemeal without any trouble, and as quickly as possible. But the removal of the body linen or flannel requires much caution, as if handled roughly the blistered or burnt skin may be torn off with it, and the accident rendered much more dangerous and difficult to treat.

Remember, that as it is always hoped the scald or burn is confined to inflaming or blistering the skin, *it is of the utmost importance not to burst the blister by tearing the skin, nor to let out the water it contains by pricking it.*

The best way to prevent either of these untoward additions to the injury already suffered, is to cut through the body-linen or flannel at a little distance from the scald or burn, and then gently lifting up to remove it, unless it stick. But if it do stick, so much only must be cut off as lifts up readily, and the sticking part of the linen must be left. The whole border of the linen or flannel covering the scald or burn must, in like manner, be lifted up where it can be most conveniently done, and as much as does not stick cut off. In scalds, generally, the whole body-garment can be removed without much difficulty,

as it is soaking with wet; but in burns, the ash
of the destroyed linen will still remain, and often
some of that part of it which has been partly
tindered will stick firmly, and must be left.

It must not be forgotten to cover up with warm
bed-clothes all the parts not injured, and to put
bottles of hot water to the feet immediately; as,
if the injury be severe, a shuddering, and even
severe shivering from cold, in addition to the
shock of the constitution, quickly comes on.   If
this continue, even for a short time, it is best
to give a little hot wine and water, or brandy
and water, or any other wholesome spirit with
hot water, to check this most distressing addition
to the patient's sufferings.

Treatment.

The object in treating scalds and burns is *to
keep up, for a time, the great heat or high temperature to which the injured part has been
raised by the scalding or burning, and to lower
this by degrees to the natural heat of the body.*
But the direct contrary is generally believed
to be what is necessary, and hence the common expression "getting the fire out," "the
fire is got out"—whereas, the fire ought to be
kept in and only allowed to go out very gradually.   This may at first sight seem a very
startling and strange statement, but it is strictly
true; and it is the principle upon which scalds

and burns are now almost invariably treated, whatever be the application.

No *cold* application ought ever to be employed ; it may allay the pain for a short time, but it will only be for a very short time ; for, as the cold application becomes warm, the pain returns and becomes intolerable. Any one can prove this for himself : if, having scalded or burnt his finger, he put it into cold water, he is almost immediately relieved from pain ; but directly the water begins to warm, or he removes his finger from it, the pain returns, and soon becomes as agonizing, or more so, than at first ; and the inflammation greater than if the cold had not been used. The necessity for keeping up the acquired heat of the part, for a time, in scalds and burns, is as great as that of keeping down the heat which a part has lost when frozen, as will be shown elsewhere.

Scalds and burns, if the blistered skin be unbroken, may be covered with dry or wet applications, whichever may be handiest or most preferred ; but if the skin be broken, wet applications, if they can be got at once, are best, otherwise dry ones must be used ; as it is of the utmost importance to protect the exposed sensitive true skin that lies beneath the scarf-skin, of which the blister consists, from the air, which renders it excessively painful.

The best and readiest dry materials are flour,

or cotton, or cotton-wadding; the wet are, spirits
of turpentine, spirits of wine or good brandy,
lime-water and oil, lime-water and milk, milk
alone, or bread and milk poultice; and all these
wet applications must be made of sufficient
warmth to feel comfortable to the finger, but not
hot.

Flour. This is best dusted on with a dredger,
and should be thickly strewed over the injured
part, and some little distance beyond it, as frequently the inflammation of the skin spreads
farther than was at first expected. After the
flour has been plentifully put on, the part should
be wrapped in a well-floured thin covering; a
cambric or thin linen handkerchief is best for
the purpose.

Cotton must be carefully pulled, so as to be
free from lumps or knots, and spread largely
and thickly over the injured part and its immediate neighbourhood, and confined with a roller.

Cotton-wadding is better than cotton, because it
lies smoother, and its glazy side renders it more
impervious to the air. The unglazed side should
be put next to the skin and three or four
thicknesses, one upon the other, and the whole
confined with a roller.

*Spirits of Turpentine or of Wine, or any other
Spirit,* heated by placing a cupful in warm
water which the finger can bear, must be gently
smeared over the injured part with the broad

web of a feather, or with a large camel's hair brush. In a house where drawing utensils are at hand, a sky brush, as it is called, is the best tool for the purpose. This anointing should be continued for ten minutes or a quarter of an hour; after which, lint or soft linen thickly spread with a liniment composed of spirits of turpentine and resin ointment, commonly called yellow basilicon, to the thickness of cream, should be applied and confined with a roller. Spirits of wine or brandy is to be applied in the same way; and may in like proportions be mixed with resin ointment to form a liniment.

The dressing should not be removed till at least twenty-four hours after, then the part washed over with proof spirit, and the same kind of liniment as before re-applied immediately. The third dressing, twenty-four hours after, should have the liniment composed of resin ointment and camphorated oil; or, if this be too sharp, calamine or zinc ointment, thickly spread on linen, may be applied and continued. This is known as Kentish's treatment, and is much practised in the coal districts.

Lime-water and linseed-oil, if the scald or burn be very extensive, is a most excellent application. One part of lime-water must be mixed with two parts of oil, and stirred round quickly with the hand. Into this a sheet or a large piece of linen is plunged and thoroughly

M

soaked, then wrung out lightly, and quickly
wrapped round the limb or body as may be.

Lime-water and milk, in equal proportions,
are to be used in the same way. Or the milk
alone may be used in like manner.

Bread and milk poultice thinly spread is also
a very good application. But in burns where
the skin has been much charred or roasted, it is
the best application from the beginning, as it
softens the burnt part.

In these various modes of first managing
scalds and burns, if the blisters have not been
broken, it is important to preserve them un-
touched for a time, as the water they contain is
a softer application, and gives less pain to the
inflamed skin, than if the blister being broken or
pricked and the water let out, the scarf-skin
drop down upon it. But even if the blister be
broken, the scarf-skin must not be removed, but
carefully smoothed down, for it still forms a
better immediate application than any other that
can be made, as it more effectually shuts out the
air, prevents evaporation from the surface of the
true skin, and thereby keeps it moist.

During the time the water remains in the
blister, nature is repairing the injury by coating
the surface of the true skin with a film of creamy
matter, which is to become the new scarf-skin ;
and it is an important point to take care that this is
not interfered with, which it will be if the blister

be left' unopened too long. Generally a few hours after the accident, unless it be very severe, the pain either ceases, or diminishes very considerably for a time. But in about thirty-six or forty-eight hours, sometimes less, sometimes more, uneasiness is again felt in the immediate neighbourhood of the blister, which soon amounts to pain on the slightest movement dragging the blister. If the part be then examined, the blister itself will look as if its contents were milky, and all round its edge a red line of inflammation will be seen. When this appears the water must be let out ; and this is best done with the point of a large needle, or by small snips with the points of a pair of scissors, in two, three, or four places, according to the size of the blister, and about half an inch above its edge. This is better than snipping it at top, as the water more readily flows out. Whilst the water flows, and the scarf-skin forming the blister drops, it should ' be gently pressed down with a little wad of lint or cotton ; and wherever any water continues lodging, the skin should be pricked or snipped, and pressed so as to get it all out, as at those parts where remaining it almost invariably causes a sore, which is excessively painful and heals last, and often with difficulty. The dressing should be simple wax and oil, plentifully spread on lint or linen ; and frequently nothing further is needed. A scald often heals, if there be no

mortification, in the course of a week or ten
days ; but a burn, as the injury is always greater,
in general is cured much more slowly. In
both cases, however, more especially in burns,
and if the scarf-skin be torn off at first, or come
off afterwards, the exposed surface of the true
skin discharges a large quantity of matter for
some time, and is very sensible. When this
happens, it should be dressed with some absorbing ointment thickly spread, as calamine cerate,
Turner's cerate as it is commonly called, or zinc
ointment. Or it may be protected by forming
an artificial crust or scab over it, by sifting flour
or starch, or powdered chalk, or calamine powder, which soon sucks up the discharge, and, as
it moistens with the continuing discharge, additional powder must be sifted on, and this repeated
two or three times a day, as may be necessary,
till a thick crust be formed. Usually there are
long cracks or chinks in the crust, through which
the discharge escapes ; but if the whole form one
single crust, and there be any uneasiness, it must
be carefully broken, so as to make an escape for
the matter. As the new skin forms the discharge ceases, and the crust generally drops off
bit by bit. But sometimes it remains, the discharge continues, and the part feels uneasy, if
not even painful ; it is best then to cover the
whole with a large wet poultice of linseed-meal,
which in the course of a few hours softens the

crust, so that it can be removed. This gives opportunity to examine the state of the sore, and the crusting may be again employed to that part which has not yet healed; or it may be dressed with either of the absorbing cerates, and rolled.

When the true skin has been destroyed, which in a scald is soon seen, by its having the appearance of soaked leather, and in a burn by its having a dried appearance like the burnt skin of roast meat, a linseed-meal poultice must be applied and continued till the dead part or slough come away and expose the new flesh beneath. This must then be dressed with calamine or zinc cerate, and rolled; or it may be treated by powdering, as already directed.

If there be no destruction of the true skin, the sore generally heals, although it may be slowly, without a scar; but if the skin be destroyed there is always a corresponding scar, and its formation is of more consequence if near the bend of a joint, as, by the natural disposition of the scar to contract, it draws the neighbouring parts of the limb together, and prevents them being again straightened.

If the scald or burn be very severe, the person's sufferings are at first so great, that it will be necessary to give some laudanum, twenty or thirty drops of which may be given as a dose to a grown-up person twice, or it may be three times a day ; but it should be brought down to

twenty, ten, or five drops, and even should be
entirely left off immediately the patient's pain
begins to diminish, and he can get sleep without
it.

REMEMBER, however, that *not one drop of
laudanum should ever be given without the advice
of the doctor*, if it be possible to obtain him.
Never give any laudanum to children; but if a
scalded or burnt child suffer very much, and be
above three years old, a teaspoonful of syrup of
poppies may be given twice or three times in the
twenty-four hours, great care being taken to
watch, lest he should become too much affected
by it, which would be known by great drowsiness, and then it must be at once discontinued.

When there is much discharge, and the wound
is long in healing, or when there is a large
slough, which is always accompanied with much
discharge, and the cure consequently very slow,
the great pull upon the constitution which necessarily follows must as far as possible be guarded
against by nourishing diet, and porter, wine, or
spirits, according to the patient's present state,
at the same time also taking into account what
his former habits have been.   Thus, when in one
case a pint or two of porter, with a glass or two
of port wine daily, would be amply sufficient for
a person who had lived moderately, yet in another, who has been accustomed to swallow beer
or gin, or both, in large quantities, at least

double, if not treble the allowance just mentioned would be necessary to keep him up at all.

During the whole treatment special attention should be paid to keep the parts as clean as possible, and to prevent the discharge soaking into the bed-clothes, by putting oiled cloth between them and the limb.

Sometimes after going on well, as it seems, for a few days, the patient begins to get dull and heavy, becomes insensible, and dies in a few hours. This not unfrequently happens with children scalded about the belly or chest; water is poured out suddenly into the belly, chest, or head, but more commonly into the latter, and destroys them.

BURNS WITH LIME.

Burns from lime are the worst form of burns, as there is always destruction of the skin and neighbouriug soft parts, proportioned in severity as to the length of time the lime remains applied. They sometimes occur from boys carelessly filling their breeches pockets with this substance, and getting wet. The lime then slakes and burns severely, quickly destroying the thin pocket, and very soon after the skin.

Any attempt to pick the lime off is useless, for it sticks fast to the skin it is destroying. The first thing to be done is to apply vinegar till the lime become changed into a harmless substance

and its burning power be destroyed. The dead
skin will require a poultice till it come away,
leaving a sore, which must be healed either by
continuing the poultice, or by dressing it with
some simple ointment and a roller, or by binding
it up with strapping.

LIME IN THE EYE.

Occasionally in slaking lime, one of the small
pieces which fly off will get into the eye, and
stick upon the front of the eyeball. It immediately produces violent pain; and if it have
fixed on the transparent part of the eyeball, a
little milky spot quickly appears, and very soon
spreads, rendering all that part of the eye white
and opake, so that the person is partially blinded.
A smart discharge of tears begins immediately
after the lime has got in, and continues. The
eyelids quickly swell and become closed, and very
severe pain follows. If but a short time pass
without anything suitable being done, the person
runs great risk of certain blindness, either from
the transparency of the eye being destroyed, or
from the front of the eye being killed by the
burn ; and when the dead part separates, the fluid
parts of the eye fall out, and the eyeball shrivels
almost to nothing.

These dreadful consequences may be prevented by at once bathing the eye with a little
weak vinegar and water, which must be applied

to the eyeball itself between the eyelids, and any little piece of lime carefully removed with the web of a feather. Whatever lime may have got entangled in the eyelashes must be carefully cleared away with a bit of soft linen, soaked in the vinegar and water. Under the most favourable circumstances, violent inflammation of the eye follows this accident, and it will be necessary to obtain medical assistance where it can be had. If, however, that cannot be obtained, leeches must be applied, and bathing with warm poppy-water or simple warm water. A smart purge must also be given.

FROST-BITE.

PERSONS exposed for a length of time to severe degrees of cold are liable to have their fingers and hands, toes and feet, nose, ears, and lips frost-bitten, and sometimes mortified ; and if the whole body be affected, the person is either frozen to death or can be restored only with the greatest difficulty.

Persons sitting still and exposed to severe cold are more liable to be frozen to death than those who are moving about ; and it rarely happens, even in this country, that a severe winter passes without a coach-guard or a waggoner being found dead in his seat at the end of the stage, or a poor wretch, who has laid himself down to rest in some entry, being picked up dead and stiff in the morning.

Those who are moving about, when their constitution begins to be affected by the cold, have an almost unconquerable disposition to lie down, and, if they do, are speedily overcome by ' that sleep which knows no waking." The instance of DR. SOLANDER, who accompanied CAPT. COOK * round the world, is known almost to every one. Whilst traversing the hills of Tierra del Fuego, he earnestly warned the party that, however weary

* First Voyage.

and tired they might feel, they should on no account give way to the disposition to sleep which would come on them, for, if yielded to for a few minutes only, it would end in death. Notwithstanding his full knowledge of the danger, he was one of the first who begged to sit down and rest, and it was only with the greatest difficulty he was compelled to move on, and thus had his life spared. Two negroes of the party, however, sank down to rest, and awoke no more.

If a frozen person be picked up before he is dead, he will be found cold and stiff, completely insensible, and his breathing suspended, from the blood on the surface being driven into the inner parts of the body and loading the brain and lungs.

Treatment of a frozen person.

It will be recollected, in speaking of scalds and burns, it was mentioned, that the object of the treatment was to sustain the acquired heat, and to lower it gradually to the common heat of the body. In restoring a frozen person, or a frost-bitten part, the object is directly the reverse, that is, *to keep the cold, which by its exposure the body has acquired, and to withdraw it by slow degrees till the body has recovered its natural heat.* If the person or part be brought suddenly into a hot room, or put in a warm bath, he or it will be killed outright. " The frozen person," says CHELIUS,* " should

---

* Handbuch der Chirurgie.

be brought into a cold room, and, after having been
undressed, covered up with snow or with cloths
dipped in ice-cold water, or he may be laid in cold
water so deeply that his mouth and nose only are
free.   When the body is somewhat thawed, there
is commonly a sort of icy crust formed around it ;
the patient must then be removed, and the body
washed with cold water mixed with a little wine or
brandy : when the limbs lose their stiffness, and
the frozen person shows signs of life, he should be
carefully dried, and put into a cold bed in a cold
room : scents, and remedies which excite sneezing,
are to be put to his nose ; air is to be carefully
blown into the lungs, if natural breathing do not
come on ; clysters of warm water with camphorated
vinegar thrown up; the throat tickled with a
feather, and cold water dashed upon the pit of the
stomach.   He must be brought by degrees into
rather warmer air, and mild perspirants, as elder
and balm tea (or weak common tea) with MIN-
DERER'S spirit, warm wine, and the like, may be
given to promote gentle perspiration."

Frost-bitten parts.

When any part of the body is first sharply attacked with cold, it becomes puffy and bluish, and
is painful and smarts.   The common expression
" Your nose is quite blue with the cold" is well
known to every one.   This depends on the blood
flowing more slowly than natural through the
vessels near the surface ; but if the exposure to

cold be still continued, the blueness disappears and the part becomes pallid or yellowish white, as if the blood were completely squeezed out of it. The pain now ceases, and the part becomes numbed and motionless, and the person is so little inconvenienced by it, that in very cold climates his first knowledge of his mishap is the accidental meeting of a friend who gives him the pleasant intelligence that his nose is frozen ; and unless some immediate means are taken to thaw it, the part is completely killed, and after due course of time is thrown off like any other mortified part. Sometimes the part is not killed outright, but seems only numbed ; as the numbness goes off, it is followed by pricking stinging heat, after which redness and swelling come on, little blisters form, and the part mortifies.

Treatment.

The principle of the treatment is in this case the same as when the whole body is frozen. The natural heat must be restored, only very gradually. Few persons have not experienced the excruciating pain which follows holding their fingers before the fire, after having been so numbed by cold as to be utterly without feeling till thus suddenly brought to the fire ; and still fewer have not been perplexed with the bitter cries of children, who, after playing with ice and snow and numbing their hands, rush to the fire

to warm them, and scream with the agony they
suffer.   Plunging the part into warm water is
most dangerous.   ASTLEY COOPER mentions the
case of a person who, having been shooting and
got his feet very cold, put them into warm water,
and both mortified.   And I myself have known a
man who, whilst sorting wet skins in very cold
weather, had his hand stiffened with the cold ; on
reaching home he put his hand into luke-warm
water, but the agony it produced was so great
that he quickly withdrew it, and could not even
bear the warmth beneath the bed-clothes during
the night.   Next day his nails were black, but
he was free from pain, and the hand was cold
and numb; he tried the warm water with the
same result, and also on the following day, but
the pain came on as at first, and the end-joints of
all the fingers and thumb had become black—in
short, were mortified.   The mortification spread,
and in the course of a week the hand was mortified
up to the wrist, and some little time after it was
necessary to cut it off.   I mention these cases to
show the importance of not suddenly attempting
to heat very cold or frozen parts.

The very cold or frozen part should therefore
either be bathed with very cold water, or rubbed
with a handful of snow, till the circulation in it
be restored ; and even after that, cold water should
be used for some time, so that the natural heat
may be recovered very slowly.

MAY follow as a tail-piece to Frost-bite, depend-ing, as their name implies, on a chill which dis-turbs the usual circulation of blood in weak parts, and is followed by a weakly inflammation. They are in general attended with much itching, tingling, and smarting, and commonly attack the heels, but sometimes even the hands of unhealthy languid persons. If paid common attention to, they usually subside in a few days; but, if neg-lected. the skin blisters, the chilblain is said to break, and very troublesome sores often follow.

Treatment.

Keeping the feet warm by wearing worsted stockings, and encouraging the circulation by rubbing once or twice a-day with soap liniment or mustard liniment, is the best mode of manag-ing chilblains in their first stage. The itching may be also for a time relieved, by brushing over the inflamed skin with dilute sulphuric acid. But when the chilblains have broken, a bread and milk or linseed-meal poultice is the best appli-cation.

SPRAINS

Consist in a straining, wrenching, or tearing of the ligaments or tough structures which bind bones together to form joints. The wrist and ankle are the joints most commonly sprained. Sprains are among the most severe accidents to which we are subject, as regards the part itself; the pain is at the moment excruciating, often returns on the slightest movement, and too frequently lays the foundation of what is commonly called White Swelling. If one therefore might have a choice of the two evils, it would be better to break a limb than sprain a joint, as the former is always the less evil of the two, in ninety-nine cases out of a hundred being cured, if the skin have not been broken, in the course of a few weeks; whilst the effects of the latter may at best remain for weeks or months as weakness or stiffness in the joint. There are few things for which medical men are more unjustly blamed than for the consequences of a sprain, and we often hear, " Why, such an one has *only* sprained his wrist or his ankle, and he has been under Mr. ——'s hands so many weeks; I wonder he does not have other advice." And so he may, and at last be like the woman who "suffered many things of many physicians, and was nothing bettered, but rather grew worse."

I shall here take the opportunity of giving a little wholesome advice to my readers. Be careful in your choice of your medical attendant ; do not choose him because he is soft-tongued and easy, and will let you do as you like. But choose him, because you are satisfied that he knows his profession ; never mind if he be a little blunt, if you be assured of his ability and honesty. Rely upon it he is quite as desirous, for his own reputation's sake, that you should be cured, and as quickly, as yourself can wish ; and therefore give him, as he deserves, your whole confidence, and do not listen to the inuendoes of any foolish tatler, male or female, who may wantonly endeavour to withdraw your trust from him. WALTER SCOTT's observation on a multitude of doctors— he might have added, a frequent change of doctors—is strictly true, when, in quoting the good old proverb, "In the multitude of counsellors there is safety," he says, "Yes, that's for the doctors, but not for the patient."

Treatment.

When a joint has been sprained it should be kept perfectly at rest, and, if the ankle, or knee, the person should lie in bed, or on a sofa. Warm moist flannels should be repeatedly applied for some hours, and a bread and water poultice on

N

going to bed. These should be continued for a few days, and no attempt made at using the joint. If the pain be very severe, and continue so for the first or following days, leeches may be applied and repeated once or oftener. Some persons are fond of putting a vinegar-poultice on at once; but I think this is better left alone till the tenderness have subsided, and there remain only a little pain and stiffness in the joint; then a vinegar-poultice is a very good application, as it produces a diversion of the low inflammation going on in the ligaments, by bringing out a crop of pustules on the skin, at a time when the pressure in rubbing any stimulating liniment cannot be borne.

When the pain has entirely ceased, the joint must not be carelessly used; and, if it be the knee or ankle sprained, walking till the joint become weak and ache must be most carefully avoided, as irreparable mischief is thereby very often set up. Short and gentle walks only therefore should be taken, and may be repeated by degrees more frequently during the day, if they do not produce pain or fatigue.

A joint often swells a long while after a sprain; under which circumstance it is best to bind it up with straps of soap-plaster or a roller.

BROKEN BONES.

Modes of removing persons injured by accident.

BY falls or blows the limbs are often broken, and such accidents, severe and painful of themselves, are frequently rendered more serious and agonizing by the awkward and careless manner in which, with the very best intentions of those who afford assistance, the sufferer is carried, with his limb dangling or rolling about, to the nearest medical man.

There is generally little difficulty, even to a not very 'cute person, in finding out whether the leg or thigh, the fore-arm or upper arm, be broken, especially if broken in or near the middle of either of these parts; because not merely is the sufferer incapable of lifting it up, but because, in any attempt to do so by himself or some other person, there is observed an unnatural bending and motion at the broken part.

Persons who break their arms, either below or above the elbow, will find it least painful to put the fore-arm at right angle with the upper, in a broad sling, which will contain it from the elbow to the points of the fingers; and if his own home or the doctor's residence be not at great distance, he will find he can walk with much less pain and shaking of the broken part than if he be moved in a carriage of any kind.

If the leg or thigh be broken, a hurdle or a shutter covered with straw, coats, or blankets, may be converted into an excellent litter, which should be laid down by the sufferer's side, and he gently and quickly lifted upon it, by just as many persons as are enough to raise him up a very little from the ground, and by no more, as, the greater number of assistants there be, the less likely are they to act together and effectually The shutter or hurdle should be carried by hand, two persons at each end taking hold of it, and all keeping step as they move along. And if a couple of poles can be procured and fixed across and beneath each end of the hurdle, the bearers will carry with less fatigue both to themselves and the patient. If neither hurdle nor shutter can be obtained, no bad shift may be made by fastening four stout poles together, and tying a blanket securely to them, so as to resemble the frame and sacking of a bedstead, and upon this

the sufferer may be laid. Hand-carriage in either of these ways is infinitely more easy than

carriage or cart, for every jolt over any irregularity in the road produces motion in the broken
bone, and correspondingly severe pain. It is
related of the celebrated surgeon PERCIVAL
POTT, that he was thrown from his horse in
Kent Street, in the Borough, and broke his
leg; some of the crowd which soon collected
about him kindly offered to get him into a
coach, but he begged to be left in the road,
where he had fallen, till a door which he pur-
chased, and a couple of chairmen with their
poles for its support, could be obtained, on which
he was carried home, as he knew that would be
the easiest mode of journeying.

Having got the person on the hurdle, shutter,
or blanket, it is a good plan to bring the sound
limb close to the broken one, and tie them both
firmly together with two or three handkerchiefs;
by doing this, great support is given to the injured limb, and any movement of it is almost
entirely prevented. And besides this, a pillow
or long pad of straw, should be placed along the
outside of the limb to render it still more steady.
In placing the limb great care should always be
taken to lay the broken bone as near as possible
in its natural direction; for if this be not attended
to, but the broken part be left bent, it is far
from improbable that one or other end of the
bone will thrust through the skin, and thereby
materially increase the mischief.

Hints for managing Broken Bones.

All broken bones should be put under medical treatment where it can be procured. But
in newly settled districts, and on shipboard, it
frequently happens that no doctor is at hand, or
likely to be met with for weeks. To meet such
emergency, some hints may be given for the
purpose of putting the sufferer in such condition as will be likely to effect a proper and useful
cure of his broken limb.

The *materials necessary* for the cure of broken
bones are few and simple, and can always without difficulty be provided, as they consist only of
linen bandages, about four fingers in breadth,
and half a dozen yards in length; of pads which
may be made of three or four layers of rug or
blanket, lightly quilted together, or pillows filled
with tow, cocoa-fibre, chaff, or leaves; and of
splints, either deal-boards four fingers wide, a
quarter of an inch thick, and of length corresponding to that of the broken limb, or wheat-
straws laid side by side to the same extent and
thickness, folded up in cloth, and quilted so as
to prevent them moving about, or even the fresh
bark of trees.

Broken limbs should not be set, as it is called,
that is, bound up with roller, splints, and pads,
for the first three or four days, as for some hours
after the accident the part continues swelling ·

if bandaged up tightly whilst this is going on, much unnecessary pain is produced, and if the bandages be not slackened, mortification may follow, which I have known to occur. It is best then, at first, only to lay the broken bone in as comfortable a posture as possible, and nearly as can be in its natural direction; and it may be lightly bound to a single splint, merely for the purpose of keeping it steady. The arm, whether broken above or below the elbow, will lie most comfortably half bent upon a pillow. The thigh or leg will rest most easily upon the outer side with the knee bent. Broken ribs and broken collar-bones are an exception to the general rule, and require immediate attention.

Broken Ribs.

A person may be presumed to have his ribs broken when, after a fall or a blow, he feels at every breath a stitch or prick in the side of his chest where he has received the injury; and if the hand be placed on this part, and the person directed to draw his breath in deeply, the broken ends of the bone will be felt moving on each other, and giving a sort of cracking feel.

Treatment.

All, in general, necessary to be done, is to wind a flannel or linen roller, about six yards

in length, and two hands' breadth in width,
tightly round the chest, so as to prevent any
motion of the ribs in breathing, which during
the cure must be performed by the diaphragm
or midriff alone. The end of the roller should
be fastened by sewing, rather than by pins,
which are liable to slip; and it is well that all
the turns of the roller should be sewn together,
which renders the bandage more secure and less
likely to require often re-rolling. If well put
on, the bandage rarely needs to be put on afresh
more than twice during the month, which it is
advisable it should be worn.

It is a common practice at once to bleed a
person who has broken one or more ribs; but it
is better left alone till the patient complain of
pain and is troubled with cough. Then, indeed,
a pint of blood may be taken with benefit, and
perhaps may need to be repeated. The bowels
should be cleared with a purge, and twenty
drops of antimonial wine, with a teaspoonful of
syrup of poppies in a glass of water, given three
or four times a day   After a few days the person will find himself much more comfortable
sitting up than lying in bed.

If the ribs be broken on both sides of the
chest, or if the breast-bone, which runs from the
bottom of the neck to the pit of the stomach, be
broken, no bandage must be applied, as it will
do mischief; but the person must be kept as

quiet as possible. These latter accidents are always very dangerous.

Broken Collar-bone.

In thin persons, there is not much difficulty in discovering this accident if the bone be broken, as it usually is, near its middle. The bump which is observed, when comparing the broken with the unbroken bone; the riding and unnatural motion felt by the fingers put on the broken part, when the arm is moved; the pain on motion; and the disappearance of the irregularity when the shoulders are brought back, and its reappearance when the hold of them is left off, are pretty good proofs of the nature of the accident.

Treatment.

This is generally easy enough, and consists in placing high up in the hollow of the arm-pit a pad about as big as two fists, and twice as wide, which must be kept in place by a tape at each end, passed one on the back, and the other on the front of the chest, and tied on a pad (to prevent galling) on the oppo--

site side of the neck. A bandage is next to be
turned once or twice round the arm immediately
above the elbow, and its two ends carried round
the chest, one before, the other behind, and tied
so as to keep the elbow close to the side. The
elbow and fore-arm are then put into a short sling,
which lifts up the shoulder, and should be tied
on the neck on the sound side. This done, all
deformity disappears, and the bone is set. The
bandages thus put on must be worn about a month.

Broken Arm above the Elbow

Is distinguished by the unnatural motion at the
broken part, and by the person being incapable
of raising either the elbow or fore-arm ; it can
scarcely fail of being discovered.

Treatment.

The pads and splints must be fitted on the
sound arm before placing on the injured limb,
and four of each will be required. The splints
should be about three fingers' breadth wide ; one
should reach from the shoulder to the bend of
the elbow ; one behind from the shoulder to the
point of the elbow ; one from the armpit to the
jutting inside of the elbow ; and one from the
shoulder to the jutting outside of the elbow.
The pads should be a little wider than the splints,
and about two inches longer, so that they may be
turned over each end of the splint, and tacked, to

prevent them slipping about. Two long rollers are also needful. The immediate swelling after the accident having subsided, the limb must be placed with the fore-arm bent at a right angle with the upper. The hand and arm are to be lightly swathed in the roller, the turns of which should overlap each other, and be continued a little above the elbow. The object of this is to prevent the swelling, which generally follows the appliance of the splints. The second roller is now to be wound round the arm twice or three times above the elbow; then the first splint is to be placed on the front of the upper arm, but not quite down to the bend of the elbow, and two or three turns of the roller made round it; next the back splint from the shoulder to the elbow placed against the arm, and the roller carried around it twice or thrice ; the third splint is now put on at the inside, its upper end being pushed up into the arm-pit, not so 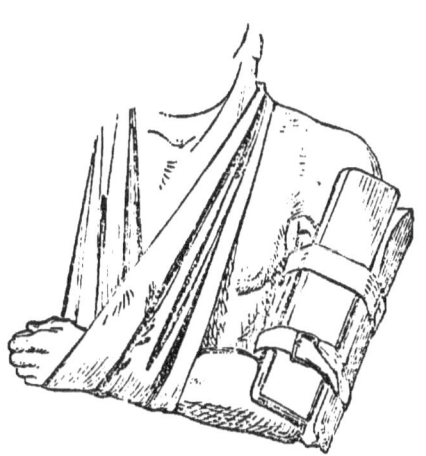 high, however, as to rub against and gall it ; and the fourth on the outside : around these the roller is now to be wound, and continued till the whole arm with the splints have been swathed from the arm-pit to the bend of the elbow.*   A short sling is then put round the neck, which must only support the hand and wrist.   By thus doing, the weight of the elbow drags down the lower end of the bone, and keeps the broken portions in place. The splints rarely require being touched for ten days or a fortnight, and must then be again applied in the same manner.   It must be worn a month or five weeks.   There is no need of keeping the person in bed, and indeed it is advisable that he should be up, as the broken bone keeps its position better than when in bed.

If wooden splints cannot be made, stiff pasteboard or wheat-straw splints are very good substitutes.   But even these may be dispensed with, and, after rolling the hand and fore-arm, a long roller well soaked in gum-water or stiff starch may be swathed round the upper arm from the elbow to the arm-pit.   The limb must then be laid carefully on a pillow, in as nearly as possible its natural position, and in the course of twelve or twenty-four hours the gum or starch dries, and a tough, unyielding, well-fitting case encloses the arm, and rarely requires being meddled with till it be completely removed at the end of a month.

---

* In some of the cuts the splints are held together by straps instead of rollers. This has been done to show the position of the splints more distinctly

Broken Arm below the Elbow.

There are two bones in the fore-arm, and, when both are broken, there is little difficulty in discovering the nature of the accident. But if only one be broken, it is not so easy for an un-practised person to distinguish it; but this is of less consequence, as the sound bone serves for a splint to keep the broken one pretty nearly in its place, even though no splints be put on.

Treatment.

If both bones be broken, two padded splints are required, extending from the tips of the fingers to the bend of the elbow in front, and to the point of the elbow behind. The fore-arm is now to be bent on the elbow: the splints applied, one before and the other behind, and both bound firmly to the arm with a roller from the fingers up to the bend of the el-bow. The arm then 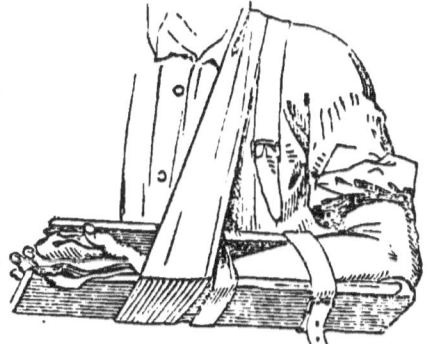 resting on its back is to be put in a sling, which shall support it from the elbow to the finger ends. The splints must be kept on about a month.

Broken Fingers.

If the first or second joint of either of the

fingers be broken, it is readily discovered; but not so easily if it be the third joint, which, however, is but rarely broken without more serious mischief.

Treatment.

A piece of thin wood or stiff pasteboard, as wide and as long as the finger, is to be placed on its front or same side as the palm of the hand. Upon this the finger being laid straight, it is to be bound with a roller an inch wide from end to end. It is best to keep the hand in a sling for three weeks or a month, and not to attempt using it till after that time. The broken finger often remains stiff a long while after it has become well knit together; it is therefore a good plan to render the joints supple by thrusting the hand for half an hour daily into warm grains, but, if these cannot be procured, soaking it for the same time in warm water; and afterwards to bend the finger gently forwards and backwards, as far as it can be moved without pain.

Broken Thigh.

If this accident occur in any part a little distant from the hip or knee-joint, it is easily ascertained by the unnatural bending at the seat of injury and by the person being unable to lift up the leg below the broken part, as well as by his not liking to attempt it on account of the pain produced by the ends of the bone pushing into the flesh.

Treatment.

Though a much more serious accident than either of those already mentioned, it may be managed quite as easily, and in many cases, if the person will be quiet, without any splints, although it is better to use them.

Without Splints.—The patient must be placed on his back upon a firm mattress, laid on a board resting on the bed-frame, which is better than on the sacking, as that sinks with the weight of the body when resting on it for some weeks. Two thick pads are to be made, of sufficient size to cover, the one the whole of the inside of the sound knee, and the other the inside of the ankle of the same limb. Both limbs must now be laid close together, in the same straight line as the body, resting on the heels, with the toes right upwards; and in doing this care must be taken that the calves of the legs rest flat on the mattress. Thus far done, the body must be kept immovable by one person, who grasps

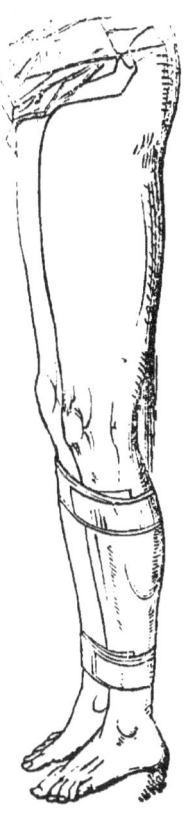

the hips with his two hands. A second person then takes hold of the broken limb with both hands, just above the ankle, and gently and steadily draws it down without disturbing its position, whilst a third places the knee-pad between the two knees, and the ankle-pad between the ankles. The gentle pulling being continued, the sound knee is brought close to that of the broken limb, but a little above it, so that it rest against the jutting inside of the joint, and then, both being kept close together, a pad about as broad as the hand must be turned round both legs, directly below both knees, and around this a roller about three yards long must be tenderly, carefully, and tightly wound, so as to prevent one knee slipping from the other. A strap and buckle will serve the same purpose; or, in want of roller and strap, a handkerchief may be passed twice round and tied, care being taken not to make the knot opposite either of the hard parts which mark the place of the two leg-bones; for if put there it will be liable to cause very uneasy pressure. Both ankles are next to be tied together in like manner, care being taken that that of the sound is above that of the broken limb. A small pad is now to be put between the insides of both feet to guard them against the pressure which is made by binding the feet together, and this completes the whole business.

This method I have employed frequently,

and, simple as it is, it is a very good one, and as good cures have followed as under the usual and more complicated mode of treatment.

With Splints.—The management of a broken thigh with splints is various, as regards both the number of splints and the position of the limb. In most cases, the straight position, with the limb resting on the heel, will answer best, and is least irksome to the patient; but sometimes it is necessary to place it on a double inclined plane, upon the summit of which the ham rests, and therefore the limb is bent, though still resting upon the heel. In all broken thighs it is specially important, to prevent inconvenient swelling, that the whole limb should be rolled, beginning from the toes, and continuing up to the hip. This requires great caution and tenderness, so as to put the patient to as little pain as possible. After having rolled the foot and leg a little above the ankle, and the body being steadied by one person, a second grasps the ankle, and gently pulls the leg down to its proper length, raising it just sufficiently from the bed, which must be assisted by a hand placed beneath the knee, and slightly raising it also, to allow the roller to be passed round it again and again till the whole limb be rolled to the hip. It is advisable that the roller for this purpose should be in lengths of four or six yards, which must be tacked together as they are used up, for a roller of fourteen or sixteen yards' length

o

is too bulky to be used with convenience to the
operator, or comfort to the patient.

If a *single splint* be used, it should be three-
eighths or half an inch thick, four fingers wide,
and of length to reach from the armpit to an inch
below the outside of the sole of the foot. It
must be measured upon the unbroken limb, and
a round hole cut, with its edge well beveled, so
as to allow the outside of the ankle to go into it,
and be saved from pressure. The whole length
of this splint is to be well padded on the side
next the outside of the broken limb. Each end
of the pad is to be turned well over the corresponding end of the splint, and then the pad
carefully stitched to the splint to prevent it
slipping about. Thus prepared, the splint is to
be put on.

The patient lies on his back on a mattress; and
the limb having been rolled, as already directed,
the body is steadied by one person, and the leg
gently pulled down, as it rests on the heel with
the toes upwards, by another who grasps the
ankle, till the sole is brought level with that of
the sound limb and there kept. The arm on
the injured side is now moved away a little from
the chest, a pad, three or four blankets or rugs
thick, put into the armpit, and into the midst of
this pad the upper end of the padded splint is
gently pushed, and there kept by a bandage,
which had been previously turned round the splint,

and tied on its outer side. The long ends of the
bandage are then passed across the chest, behind
and before, crossed on the oppo-
site side, brought back again,
and tied upon the splint.
Another bandage, fastened to
the splint in the same way, is in
like manner to be passed round
the hips, and tied also on the
outside of the splint.

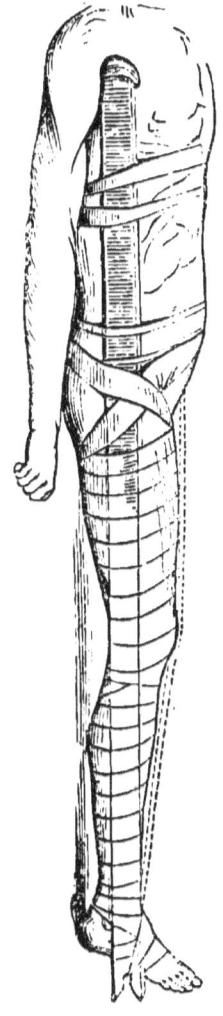

It now remains to fasten the
broken limb to the splint, which
is done with a roller four fingers
wide, and about sixteen yards
long ; and this also should be in
shorter lengths, which are to be
tacked together as they are used
up. The outer side of the limb
is first gently brought close to
the splint, and the ankle having
been well fitted into the hole
made for it, the limb and splint
are held firmly together by the
hands of one person on the
thigh, one hand above and the
other below the broken part,
and the leg also by another
person, who grasps it and the splint just below
the knee, whilst the person pulling at the ankle
grasps it and the splint together, still continuing

to draw. The person who puts on the bandage
now passes it two or three times round the
foot across the instep, upon which it is to be
carefully tacked through all the turns. This done,
the bandage is passed over the splint, and round the
ankle two or three times, then again down under
the sole of the foot, into the fork of the splint,
across the instep, round the ankle again, over the
instep, under the foot and the fork of the splint,
and again round the ankle, so that in this way
the bandage forms a figure of eight from the leg
to the foot, the crossing of which is on the front
of the ankle. Great care is required in putting
on this part of the bandage, as upon it rests the
whole scheme of the treatment, which consists in
preventing the lower end of the broken bone
being pulled up over the upper end. The top
end of the splint resting in the armpit being there
fixed, the intention is to keep the foot and ankle
fast to its lower end, and thus preserve the position of the lower end of the broken bone against
its upper end. When this has been done, it only
remains to bind the leg and thigh to the splint
by carrying the roller up again and again over
the limb and splint, each succeeding turn of the
roller slightly overlapping the foregoing one,
till the hip be reached, and then three or four
turns are made round the splint and hips, and
the finish put to the whole by tacking the bandage
firmly together. The limb is now gently laid

down upon the mattress, with the toes upwards; and to prevent the foot lolling to either side, which would disturb the position of the broken bone, a bandage should be passed once or twice round the ankle, its ends crossed upon the instep, passed once or twice round the foot, tied on the instep, and then its ends fastened one to each of the curtain-rails on the sides of the bed.

If this bandaging be properly managed, and the person keep tolerably quiet, it will not require to be meddled with till after a fortnight or three weeks, when it must be reapplied as before ; but it will be advisable only to remove one length of the roller at a time, and replace it by a fresh one before taking off the others. This will have the advantage of keeping the limb steady, and not requiring so many assistants. It will also be proper in taking off that part of the roller which fastens the foot to the lower end of the splint, that the foot should be held in its place, so that it neither loll to either side, nor be otherwise disturbed.

Sometimes it happens that, for the first few days after the broken limb has been set, there will be spasm in the thigh, which pulls up the lower broken end over the upper, and by thrusting its sharp points into the soft parts, keeps up the spasm. In addition to the pain thus caused, the limb will be shortened, a circumstance, however, which may occur also without any painful

spasm. When this takes place, it must be prevented by weighting the foot sufficiently, which
is easily done by passing a bandage once or twice
round the ankle, bringing its ends across the instep to the sole of the foot, and slinging a brick
or a seven-pound weight, which must hang over
the bed-foot, to which a bit of board, about four
inches high, should be screwed, so as to form a
pulley on which the bandage may run and play.
Generally the need for the weight ceases after
three or four days, the muscles having then become tired ; and so soon as this is the case, the
weight may be removed.

If the accident happen at sea, or the person
have to be moved from place to place, and liable
to be shaken, it will be best

To use four Splints.—The principal splint is
the outer one, which must be of the same length,
and fastened to the body, and to the foot and
ankle in the manner already mentioned ; but the
whole limb is not to be bandaged up till the other
splints are put on. One splint should be placed
on the inside of the limb, and must reach from the
fork of the thighs to an inch below the inside
of the sole of the foot, with a round hole cut
in it to receive the inside of the ankle. Its
upper end should be tied first with a handkerchief round the upper part of the thigh, to keep
it steady, and afterwards the lower end fastened
to the ankle and foot, and to the outer splint,

with the roller which had already been begun to be used. Another splint should now be put at the back of the limb, reaching from the share- or sitting-bone, which may easily be felt just where the buttock joins the top of the thigh, to about two inches above the heel, and this lower end of the splint should be hollowed a little, so as not to dig into the skin. Two or three turns of the roller will steady this, and then the last splint must be put on in front. This front splint should reach from about an inch below the crease, which separates the bottom of the belly from the top of the thigh, to an inch above the bend of the ankle. At the part where this splint will lie upon the knee-cap, three or four saw-tracks must be made across it, about half an inch apart, and nearly through its thickness, so that the splint will bow here, otherwise the pressure it makes upon the knee-cap will be unbearable. This splint having now been put on the front of the limb, the roller is to be continued round, and run up to the top of the thigh, covering all four splints at the same time. In this way the limb will be enclosed in a long box, and it is hardly possible, without vio-lence, to displace it.

Great care must be taken to inquire constantly during the progress of the cure whether the splints pinch or wring any particular part; and more commonly the ankles are the parts which become so annoyed. Wherever the person com-

plains of this, the bandage should be cut through
a little above or below, and several turns of it
having been taken off, some lint or other padding
must be gently pushed in to relieve it, and then
the roller replaced, and carefully sewed together
where it had been cut through.

It will be necessary that either of these splint-
ings should be continued for at least six weeks ;
and if, at the end of that time, on taking the
splints off, the person cannot raise his leg a little
clear of the bed, and, more especially, if the
thigh be noticed to bend at the broken part, the
union is not perfect, and they must be put on
again for three or four weeks more ; but this is
not often needed.

Sometimes, though rarely, this straight posture
cannot be borne, and it is necessary

*To place the limb, with the knee bent, over a
double inclined plane.*   This double inclined plane
is very simple, and easily made, consisting of two
boards half an inch thick and two feet wide ; one
should reach from the sitting-bone to the ham,
and the other from the ham to an inch below the
heel.   They are then to be joined endways in
such manner as to form an angle, the ridge of
which should be about six inches above the other
ends of the boards, and prevented splaying by
one or two braces at bottom.   Some pegs are
usually dropped into holes on each side of the
broken limb, to prevent it slipping about.

The broken thigh is now to be brought close to the sound one, and the knees and ankles

having been tied together with handkerchiefs, the knees are to be gently bent, the heels a little raised, and the inclined plane entirely covered with a large pad, six or eight folds of blanket thick, carefully pushed beneath them; which done, the limbs are gently dropped upon the plane.

The further bandaging may be either simply tying the knees and ankles together with a pad between them, as first described (p. 190), or

Three short splints may be put on—an outer one, extending from the top of the outside of the thigh to the outside of the knee; an inner one, from the fork of the thighs to the inside of the knee; and a front one, from a little below the crease of the groin to a little above the knee-cap. Three bandages or straps, guarded with a pad each, must be gently pushed beneath the thigh, where the pads are left to prevent cutting; and these ends of the bandages being brought out on the opposite side of the broken thigh, are tied each to its other end over the splints at the upper, lower, and middle parts of the thigh, as tightly as can be borne without pain.

Broken Knee-cap.

This accident very often happens, sometimes from falling upon it, but more frequently by the effort made to prevent falling, in making a false step on the stairs, or in slipping off the kerbstone, and immediately it is thus produced the person drops like a shot, and when lifted up cannot stand on the limb of which the knee-cap is broken.

When, after such a fall or slip, the person is incapable of bearing on that limb, and neither thigh nor leg be broken, and the movements of the hip, knee, and ankle are undisturbed, the knee is to be carefully looked at and felt. If this be done very soon after, and before much swelling comes on, there will be found, instead of the cap of the knee, a pit on the front of the joint about an inch and a half long, into which the fingers immediately drop with the least pressure, and above and below the pit will be found a bone, neither of which is so large as the knee-cap of the sound side, but much more moveable than it. These are, in fact, the two pieces into which the bone is generally broken.

Treatment.

This is very simple; though very rarely indeed, under the very best management, do the broken pieces unite properly, that is, by bone.

The person must be put on his back in bed, with his head and body raised so as to be in a half-sitting posture. The thigh and leg are to be kept in the same straight line, and the foot and leg raised as high as can be conveniently borne, so that the whole limb bend upon the body at the hip-joint. In this posture the patient is to be kept by a short sling, the upper part of which passes round his neck, and the lower round his foot and heel ; in fact, to use a common expression, he is "tied neck and heel together." In this way only can the broken pieces of bone be brought at all near together, for the muscles of the thigh pull up the upper piece and prevent it being drawn down, whilst the lower piece is so fixed to the shin-bone that it cannot move without moving that bone. The upper end of the bone is, therefore, left alone ; but, by bending the limb on the belly, the lower piece is brought up to or near it, and there kept by the sling. After the swelling, which is often very great, has gone down, generally at the end of a week, it is the common practice to put on one circular strap, or two or three turns of a roller upon the thigh, immediately above where the upper piece of bone is felt, and sufficiently tight to prevent it slipping under. Another circular strap or roller is put in like manner upon the leg directly beneath the lower end. A couple of handkerchiefs tied round these parts will answer the same purpose. The two circular bandages are now brought together, the upper one drawing down with it the upper piece of bone a little, by tapes from one to the other, and tied on each side of the knee. This posture and bandaging require to be kept up about a month, when they may be discontinued, as after that time little more benefit is obtained.

When the person first gets up, he is not very well able to bend his knee, but this is soon got over. Yet he still finds that his knee is very weak, his leg unable to support his weight, and that it cannot be thrown forward with steadiness and safety in stepping out. This arises from the substance by which the broken bone is united stretching; and if this stretching be great, as it occasionally is to several inches, the person becomes quite lame and incapable of standing, in consequence of the muscles which brace the leg to the thigh becoming lax by the lengthening of the new substance allowing the upper part of the knee-cap to which they are fixed to rise above its proper place. This laxity of the muscles is overcome by compelling them to shorten themselves as much as necessary by a very simple proceeding. The person must sit upon a high table with his leg hanging over, just clear of the knee, and then must swing it backwards and forwards till he can raise it straight with his thigh. When able to do this, he must fasten a pound or two weight to his foot, and proceed as before. After which the weight is to be increased once or twice. A week or fortnight's practice in this way will put the muscles to rights, enable them to brace the knee properly, keep it straight to support the body, and also throw the leg forward so as to render the person capable of walking safely.

Broken Leg.

The leg has in it two bones. The great bone of the leg, or shin-bone, is very distinct, and can be easily felt from the knee to the ankle-joint, being little covered with soft parts, and its sharp front edge is commonly called the shin. The little bone lies on its outer side, but being well covered by flesh, cannot readily be felt, except at the outside of the ankle and a little way above it.

When both these bones are broken, at almost any part, except quite close to the knee or ankle-joint, the accident is easily discovered by the

unnatural motion at the broken part, and generally by the regular line of the skin being there
interrupted by a projection.

But one only of these bones may be broken,
whilst the other remains unhurt. Now if the
little bone be broken, although the person may
feel great tenderness when the broken part is
pressed on, and may have pain when he attempts
to stand or walk, yet it will, in general, be almost impossible for any one but a medical man
to determine certainly that the bone be broken.
But fortunately it is not of great consequence,
for if the patient only lie still, as it is most probable he will do, because he suffers pain in
attempting to move about, the broken bone will
unite very satisfactorily, because the shin-bone
itself makes the best splint that can be provided,
and keeps the little bone in place.

But if the great bone be alone broken, there
is generally irregularity at the broken part, and
a little movement may be there felt, so that the
accident can be found out. Even in this case
the other bone serves as a good splint, and if the
person only keep the limb rested for a time,
there will be no positive need of splint or bandage,
although it would be best to put them on.

Treatment.

In most cases a broken leg can be managed
very easily and well, by merely rolling it, from

the middle of the foot to the knee, in a long
bandage well soaked in thick starch or gum-
water, of which the latter is best.

It will be necessary to wait till four or five
days after the accident, by which time all the
swelling will have gone down. During this
time the leg should be laid on its outside, upon
a pillow, with the toes a little raised by a pad
placed beneath the outside of the foot near the
little toe, and the knee should be half bent.

Before putting on the roller, the foot and leg
must be wrapped smoothly in a double fold of
lint, otherwise the gum will stick to the hairs,
and there will be much difficulty in getting the
roller off afterwards. This done, the leg must
be gently raised, and supported by two persons,
one of whom holds it above the broken part, and
the other below, with one hand round the ankle,
by which a little pull is to be made, so as to
prevent the broken ends of the bone overlapping.
The roller is then to be put on, turning it first
round the middle of the foot, and continuing it
over the instep and heel on to the leg and up to
the knee, taking care that each turn of the roller
half cover the one just made. Having reached
the knee, the roller must be turned round the leg
in the same way downwards to the middle of the
foot, and again upwards to the knee, and there
left. The limb is then laid down on its outside
upon a smooth pillow as before, and the front

oi the foot supported at such height that the tip of the great toe and the knee-cap are on the same level. Care also must be taken that the leg should be put as nearly as possible in the same direction as it would lie if unbroken. In the course of twenty-four or six and thirty hours the roller will have dried, and a firm close-fitting case is formed, in which the leg will be immoveable. When the bandage is hard and firm, usually about the third day, the person may get up and move about, and it is not uncommon to see him hobbling along with the aid of a stick without much inconvenience. Sometimes it may be necessary to take the bandage off and re-roll it, if pinching anywhere, or if, by shrinking of the soft parts, it get very loose, but usually it does not require to be meddled with till the end of the month, when it may be entirely removed.

If splints be used, two are required, three or four fingers in width, according to the size of the leg, and reaching from the knee to the sole of the foot; each having a circular hole cut out where they will rest against the ankle. The splints having been thickly padded, the leg, placed as already directed with the knee bent, is to be gently raised, and one splint slipped beneath it along the outside of the leg; the other is laid upon the inside, and then both are fixed by winding a roller around them from the foot to the knee.

The leg resting on the outside, with the knee bent, is generally the best and easiest position. But sometimes the broken ends of the bones will not drop into their proper place, or will not so remain when the leg is thus laid. It then becomes necessary to put the limb straight and resting on the heel, and if there be still any disposition in the broken ends of the bone to stick up, it will be necessary to weight the foot, as directed in the treatment of broken thigh, for a few days,  till the disposition of the muscles to drag up the lower part of the bone ceases.

Broken Toes

Rarely occur without severe injury of the soft parts, and excepting in the first joints of the great toe and that next to it, can only be discovered with difficulty.

Treatment.

A piece of thick pasteboard may be placed on the under surface of the toe and fastened to it

P

with a few turns of a narrow roller. But if the
person be kept quiet on a bed or sofa, there is
not actual necessity for splint or roller.

BROKEN BONES WITH WOUNDS OF THE SOFT PARTS RUNNING DOWN TO THEM, OR COMPOUND FRACTURES.

These are very serious accidents on many accounts, and require careful surgical assistance,
where it can be obtained. It will not, however,
be useless to give a few general hints regarding
them, in case no proper aid can be procured.

A compound fracture is serious in proportion
to the size of the wound, and the tearing and
bruising of the soft parts; the more severe this
part of the injury is, the more dangerous is the
accident. A compound fracture is most dangerous when a joint is involved in it. It is more
serious in the lower than in the upper limbs; is
more to be dreaded in the thigh than in the leg,
and more in the arm above the elbow than below it. The great object in the

Treatment is to make the accident a simple
fracture by healing the wound as quickly as
possible, which, in the thigh especially, is a
very difficult business. In all cases it must
be at first attempted to unite the edges of the
wound by bringing them lightly together with
strips of sticking-plaster; and the limb should
be covered with a light, cold, wet linen cloth,

which must be repeatedly moistened by squeezing a wet sponge over it, or by sprinkling it with water, as, by evaporation, it becomes dry. The object of this is to regulate the inflammation which less or more severe generally ensues. That the evaporation may be kept up, and the limb thereby cooled, it will be necessary to keep the bed-clothes away from it by putting a cradle across, over which the sheet alone should lie, care being taken at the same time that the edge of the sheet should be lifted up in two or three places, so that there may be a current of air; otherwise the limb will be kept in a steam-bath, and damaged rather than relieved. The use of a cradle is necessary only for the thigh or leg. The arm can lie on a pillow uncovered by the bed-clothes ; and if a cradle be used, it is merely for protection from injury.

If happily the wound heal soon, much of the dreaded danger ceases, and after a few days have passed by, the accident is to be treated precisely as if there had been no wound. But unfortunately this is not of frequent occurrence.

In general, after three or four days, the patient begins to get fidgety, cannot sleep, or only gets short and disturbed sleeps. He soon begins to be hot and thirsty ; his head aches, he becomes more restless, has one or more shivering fits, and usually becomes more ill towards evening ; his mind wanders, or he even becomes delirious, and

dies, in the course of ten days or a fortnight, from
the violence of the constitutional disturbance
caused by the injury. Under more favourable
circumstances, and with these symptoms less
severe, the wound begins to discharge at first a
dirty bloody sort of matter in small quantity,
which by degrees increases, and if things go on
well changes its character to that of good matter,
which is free from smell, about as thick as cream,
and of a straw colour. With the appearance of
such matter the symptoms mentioned soon subside, the fever goes off, the sleep and appetite
return, and then begins the second contest between the constitution and the wound, which not
having united at first, has a long process, in surgical language called, *union by granulation*, that
is, the formation of new flesh to fill up the gap
formed by the injury, to pass through, before the
broken ends of the bone can begin to knit together. This is a very perilous stage in the
cure of the accident for persons whose health
has been broken by intemperance, age, or any
other cause ; and if the injury have been to the
lower limb, they most commonly die, unless the
limb be cut off; and even this is a very uncer-
tain remedy. Country people generally do better
under compound fractures than those living in
towns ; and children better than grown up persons, indeed the very severe accidents which
children will scramble through are so astonish-

ing, that with them there is always hope under circumstances which with adults would not hold out the least expectation of a favourable result. If the constitution fail in this second stage, the feverish condition again sets in, the pulse becomes quick and weak, the countenance is flushed with pink, and alternate heat and violent perspiration, general wasting of the body, loss of appetite, dry brown tongue and restlessness, are soon followed by the person beginning to wander, and then becoming delirious, and death closes the scene. When these symptoms set in, the wound ceases to discharge, or discharges only a thin watery stinking matter, and has the appearance of being glazed, and not unfrequently the skin and neighbouring soft parts mortify, and if there be strength of constitution to throw off these dead parts, the broken ends of the bone are exposed, generally dead, bare, and of a whitish colour.

Directly the constitutional disturbance begins, the wound must be poulticed, to encourage the formation of matter, as its appearance and production, of a good sort, is, as has been mentioned, a very favourable symptom. The poultice must also be continued till the wound have filled with new flesh to the surface, and indeed that is the best application till it have nearly or entirely healed.

The *Medical treatment* differs in the two stages of the constitutional disturbance. In the *first stage*, when the inflammatory condition is accom-

panied with strength, it will require checking with
occasional doses of calomel and tartarised antimony, which however must be employed with
great discretion, as not unfrequently, and if the
case go on badly, after three or four days the
symptoms assume a typhus-like character, and
instead of depressing the constitution, it will
require support with wine and other stimulants,
or the patient sinks at once.

In the *second stage* the inflammatory stage is
of tnat kind depending on exhaustion, and then
a once the constitution requires to be assisted
by everything which will prop up and strengthen
it ; wine, brandy, and strong nourishing broth,
or nourishing easily-digested food must be given
often in very considerable quantities.

I am aware that in cases of compound fracture, when these severe symptoms set in, it can
scarcely be hoped that a person who has not
received a surgical education will be likely to
be able to carry a case through successfully.
But still writing, as I now am, for circumstances
under which no proper medical aid can be obtained, it does not seem out of place even to
notice such serious cases, and to lay down a few
general rules which may enable an intelligent
person to render assistance which might possibly
save even one valuable life.

I cannot conclude this subject of Broken
Limbs without mentioning the following history,

which used to be related by one of the HUN-
TERS in his Lectures, as it affords encouragement for an unprofessional person making attempts to treat a broken limb when no doctor
can be procured, and also shows how well Nature
can manage when left to herself. A madman
at Edinburgh, being sometimes sensible, had the
privilege of walking in the garden of the mad-
house with a keeper. He one day attempted
making his escape whilst the keeper was at a
short distance, and getting to the top of the
wall, which was but a little height from the
ground on the side next the garden, jumped
down on the other, where the ground being
much lower, his fall occasioned a compound
fracture of the leg. He was carried to the in-
firmary, the fracture reduced, and secured by
the eighteen-tailed bandage and splints. He
was very unruly all the time the surgeons were
engaged in setting the limb; but as he seemed
pacified afterwards, they left him, hoping he
might get some sleep. As soon as they had
withdrawn, he very carefully took off the splints,
bandages, &c., and placed them in the same
manner on the sound leg. Then tearing a hole
in the ticken of the bed, he thrust the fractured
leg among the feathers. When the surgeons
came next day and took off the bandage, they
were surprised at not finding any fracture. The
physician who was present, as in those days it

was customary for the physician and surgeon to
visit together, asked the surgeon how this had happened.   The surgeon replied, he could not tell ;
he was certain there was a fracture the day
before.   At this the madman was very angry :
" Pretty fellows," he said, " not to know when a
leg was broken ; but to bind up a sound leg for
a broken one ! "   The doctors then insisted on
seeing the other leg, which the patient said was
very well, and pulling it out from the bed, shook
it at them, saying, " See this is a sound leg."
Upon examination they found the feathers had
become so clogged to it by the blood, as to keep
the broken bone in place and admit of his
stirring his limb about.   As the tension seemed
nowise increased, they thought it best to humour
him, and let the leg remain as it was, as he
possibly would undo all their work.   So to
please him they bound up the sound limb again
with bandages, splints, &c., leaving the other to
itself, and it did perfectly well, the feathers not
falling off before the bones were consolidated,
and no ill accident resulted.   The whole matter
is readily explained : the clotted feathers formed
as complete a cast as if plaster of Paris had been
used, a mode of treatment sometimes required
when a troublesome patient cannot be managed
in any other manner, by which the broken bone
is kept perfectly steady, and thus is in the best
condition for union.

The *cradle* to which reference has been once or twice made, consists of some curved iron wires passed through three wooden laths as in fig. A. But one may be easily made by cutting a wash-tub hoop in two or three pieces, and nailing it to two laths as in fig. B.

A   B

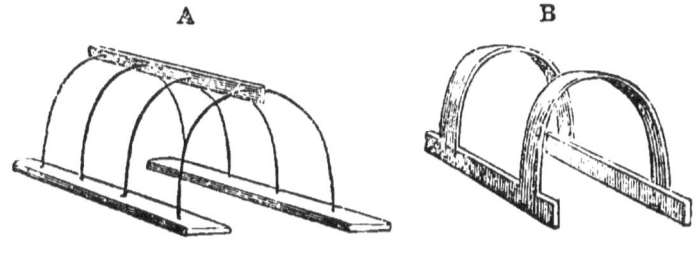

BENT BONES.

WEAKLY children, or such as are disposed to be rickety, occasionally in falling, bend their bones, instead of breaking them. This accident more commonly happens in the fore-arm than else-where, giving it a very distorted appearance, but is not attended with much, if any enduring pain.

In such cases it will not do to bend the bones back suddenly to their proper position; for if this be done, they will very probably be broken in the attempt. But a splint must be placed on the concave side of the arm, to which the arm must be bound gradually tighter from day to day till its straightness be restored.

DISLOCATIONS.

When a limb or part of a limb slips out of its socket or joint, it is said to be dislocated, or, in common language, "put out." It would be useless here to attempt giving any, even general description of the signs or appearances by which the various dislocations of the several joints are made known to the medical man, more especially as some of them are not only very difficult of distinction, but equally difficult in being replaced.

There are, however, a few dislocations, which having once happened, occur again and again, and more readily in proportion to their frequency, so that the unlucky subject of the accident is pretty well aware of what has happened to him, and if he have had it put to rights without much tackle, may be able to give tolerable direction to an unprofessional person who may be disposed to render him assistance.

The dislocations which most frequently recur, and with slight effort in the person who has once suffered them, are dislocation of the jaw, and dislocation of the arm into the armpit, and, less commonly, dislocation of the thigh at the hip joint.

Dislocation of the Jaw

may happen on cne or both sides, more commonly the latter. It mostly takes place in gaping, when the lower jaw being violently and quickly drawn down, its joint-ends slip from their sockets, and the jaw becomes firmly fixed, keeping the mouth wide open. The face in consequence is lengthened considerably; the expression altered and vacant; the power of speaking lost; and any attempt at utterance producing only strange and incomprehensible noises and the oddest contortions of the countenance possible, and often rendered exceedingly ludicrous by the various shifts the person employs in endeavouring to make himself understood. An amusing illustration of this accident was used to be enacted by ABERNETHY with great humour. An officer was dining with a party of friends, and his laughing faculties having been excited, he was rattling along and laughing heartily, when suddenly he became dumb, or rather, he ceased to be able to speak, his mouth remained wide open, and he uttered only a vast variety of strange sounds. At first it was supposed he was endeavouring to amuse the company by these uncouth noises; but soon it was perceived to be no joke, and that he was really unable to close his mouth, or to speak. After a little while he managed to make them understand he had dislocated his jaw, and that it

would be necessary to send for a doctor, who in
due time arrived, and set about replacing the
jaw. But whether it was, he did not know how
to perform the operation, or whether he put in
one side, and whilst attempting to put in the
other, the former slipped out again, as it will
sometimes do, he could not manage the job at
all, and the officer, who had frequently suffered
from the same accident before, and had it replaced without difficulty, getting angry, and at
last furious at his bungling, induced the doctor
to change his tack, and declare the sufferer was
mad. This of course alarmed the whole party,
who seized on the unfortunate soldier, carried
him to bed, and put him in a strait-waistcoat,
whilst the doctor prepared for shaving his head
and putting on a blister. The poor fellow
finding by this time he could not hope by further exertions to make his condition understood,
or free himself from his tormentors, and the
doctor still persisting he was mad, at last
made signs for pens and paper, which as it was
thought he could do no mischief with, and that
his asking for them was rather a sign of returning reason, they were brought, and he immediately wrote, "For goodness' sake send for
Mr ——, the surgeon of my regiment; he
knows what's the matter with me." The letter
was dispatched, the surgeon soon arrived, the
dislocation was quickly put to rights, and the

ignorant blockhead who had caused all the tur-
moil slunk off in disgrace.

The *reduction or replacing* of a dislocated jaw,
either on one or both sides, is very easily managed. The patient being seated on the floor,
and his head resting against the operator's knees,
who stands behind him, a couple of fork-handles,
or two pieces of hard wood about the same size,
are to be passed into the mouth, one at each corner, and to be pressed back as far as they will go
between the back teeth on each side, and there
held by another person. The operator then
bending over the patient, and passing his own
fingers between one another so as to make a loop
of both hands, places
them under the chin, and
pulls it up so as to close
the mouth. As this is
doing, the joint-ends of
the jawbone are made to
descend, and as soon as
they reach the edge of
their sockets, are pulled
into place, and the dis-
location is reduced. Care
must be taken that the
pulling up of the chin be
made level, and that the

fork-handles both retain their place, otherwise if
it be unequal, or one of the forks slip, only one

side of the jaw goes in, and very commonly in
attempting to reduce the other, it slips out again :
and this often is repeated several times to the
equal vexation of patient and doctor. When
this accident occurs the first time, it is well to
keep the jaw closed for two or three days, by
passing a bandage once or twice round the top
of the head and under the chin; and the person
should be advised to be cautious how he laugh
or yawn too widely, as when the jaw has once
slipped out, it readily does so again in either of
these actions. Afterwards, however, it is little
needful to put on any bandage.

Dislocation of the Arm into the Armpit.

Although this dislocation is generally produced
by falling on and suddenly jerking up the elbow
when thrown a little way from the side by the
falling person in attempt to save himself, yet it
may occur in children from other causes, of which
it is advisable to give some warning. Thus it
may be produced in play, by the frequent and
foolish practice of catching hold of a child's hands
and wrists, jerking him off his feet and swinging
him round in a circle, the performer of this
silly practice himself forming ‘ the centre
around which the child spun. Catching hold of
a child's hand and suddenly jerking him over a
channel or dirty place in the footway may also

dislocate his shoulder.   These two thoughless proceedings should therefore be avoided.

When the arm has been once dislocated, it also is very liable to be dislocated again and again, by any sudden jerk of the elbow upwards, or even in the simple effort of putting on the coat. The person soon becomes well aware of the injury, and finding he can neither get his elbow close to his side, nor raise it to a level with his shoulder, is pretty sure he has "put his shoulder out," as he had done before.   In general, the more frequently the bone is put out, the more readily it is put in, if only managed in the right way, and shortly after its occurrence.

Mode of Reduction.

The patient and the person who is to pull the arm into place, both lie down on a sofa, or still better, upon the floor on their backs, side by side, but in contrary directions, so that the feet of the one are at the shoulder of the other, on the side where the displacement is. The operator then having taken off his shoe and put a folded towel in the patient's armpit, puts his foot upon it between the chest and the arm, using the right foot if the right shoulder be dislocated, and the left, if the left shoulder.   He then grasps the patient's wrist with both hands, and pulls the arm down steadily.   At the same time he tells the patient to make some little change in his position, and thus inducing him to call some other muscles into action, the resistance to the reduction, which the muscles of the dislocated shoulder had been previously offering, is for a moment suspended, and at that moment the operator pulls a little more vigorously, and generally the bone immediately returns to its socket with a more or less loud snap. This momentary diversion of the patient's attention greatly assists in the reduction, and if adroitly acted on, will often enable a person not very strong to reduce the dislocation in a man of twice or thrice his own power, and who without it will often resist effectually

the conjoint assistance of three or four stout persons who may be tugging at his arm with all their might.

A person who has repeatedly dislocated his shoulder, may, if he have courage to bear a little pain for a few minutes, even manage himself to reduce it, if the accident have happened whilst he is out in the fields, and there be a five-barred gate at hand.   All he has need to do is to get his arm over the top rail, and then, hav-

ing grasped the lowest rail he can reach, hold fast, and let the whole weight of his body hang on the other side of the gate, and then, if he make some little attempt to change the position of his body, still, however, letting its weight tell on the top of the gate, the bone will probably slip into its place.   The principle on which this is done is exactly the same as when the heel is put in the arm-pit and the arm pulled, that is, to move the head or top of the arm-bone to the edge of its socket, below which, when dislocated, it had dropped, and this done, the muscles of their own accord pull it into place.

Dislocation of the Thigh at the Hip-joint does not so frequently recur after one displace-

Q

ment as the dislocation of the shoulder just mentioned; but I have had one patient in whom it
had been dislocated nineteen times.   Persons

who have had disloca-
tion of the hip two or
three times are pretty
much as well aware
of it as those whose
shoulder has been put
out.   Still, however, it
is much more likely
to be confounded with
some other accident
than at the shoulder.

Mode of Reduction.

If there be tolerable
reason for believing that
the thigh is really dis-
located, its reduction
may be attempted in
the same way as in dis-
located shoulder.   At
any rate an attempt is
worth trial.   The pa-
tient and the operator
both lie down on their
backs; and assistants
hold the hips of the
former steady, so that they shall not sway about.

The operator then puts his leg between the patient's legs and presses his foot close up to the fork, which must be protected with a towel; he then grasps the patient's ankle with both hands, and pulls; bids the patient change his position a little, and, whilst he is thus engaged, pulls a little more briskly, and probably succeeds in replacing the bone, which goes in with a snap.

Dislocation of the Neck

Is a most serious matter, in general, either destroying the person on the spot, when he is said to have "broken his neck," or causing complete palsy of every part below the injury, so that the poor wretch is like the king in the Arabian Nights, who was changed into marble with the exception of his head, and dies in a few hours. Such cases are of course beyond either Household or Professional Surgery.

But there is another form of this accident as certainly fatal, though not so marked by its symptoms, nor destroying life so quickly. It is, however, right to direct attention to it, for the purpose of warning against the silly practice of showing a child "the way to London," which, as most people know, consists in putting one hand beneath his chin, and the other beneath the back of his head, and thus lifting him off the ground. The consequence of this folly is, occasionally, and especially if the child be heavy,

to tear asunder the ligaments which bind together
the skull and uppermost bones of the spine, and
then the head falling forward, the spinal marrow
is squeezed, and the child killed. ASTLEY
COOPER used to mention an instance of this kind,
in which the child's constitution seeming aware
of the mischief done, the boy walked gently
about for a time, most carefully steadying and
balancing his head, as it had been a full vessel
of water, and if another child accidentally ran
against and jostled him, he was obliged to get
to the nearest post, or other support, upon which
he planted his elbows, and with his arms upright,
rested his chin upon both outspread hands, till
he recovered the shock he had received. Nothing
can be done for this accident, and the child
lives only, with the probability of his life being
destroyed, at any time, instantaneously. I trust
that thus showing the great danger of this very
silly play, will prevent any one who has hitherto
practised, from repeating it, after this warning.

It may be not quite out of place to notice here

The so-called " Dislocation of the Neck."

It may be occasionally heard among sporting
people—" Mr. —— was thrown from his horse
and dislocated his neck, but it was put in directly,
and he rode on as if nothing had happened."
Now, in such case, there is no dislocation of the
neck, which, as just shown, is a very dangerous

accident, and any attempt to replace the dislocation after the huntsman mode, would kill the
person outright.

In this hunting accident all that happens is,
that the person thrown from his horse pitches on
one side of his head, which forcibly twists his
neck in the opposite direction, and the muscles
on the struck side, being violently wrenched,
cannot recover themselves, and consequently the
head remains fixed with the face looking incon-
tinently over the opposite shoulder. The remedy
consists in seating the man on the ground, and
placing his shoulders between the knees of
another person, who lays hold of the head with
both hands, gives it a twist in the other direction,
and all comes right. I have known this accident
happen to a man who fell from a ladder, and
relieved in precisely the same way.

RUPTURES.

LABOURING persons are more particularly liable to this complaint, which is produced by a portion of one of the bowels slipping out of the belly, generally after great exertion, as dragging or lifting heavy weights, jumping, and the like; though sometimes children are born with the disease.

The most usual seats of rupture are in the groin and its immediate neighbourhood, and at the navel. The latter kind is very common in young children who scream violently, and in common language it is said that "the navel has started."

Ruptures which exist at birth, or which are produced in early life, can generally, by care and attention, be cured before manhood; more especially if at the navel.

If happening after birth, a rupture generally shows itself as a swelling suddenly appearing in the groin after violent exertion; remaining distinct whilst the person stands upright, disappearing when he lies down, and returning again when he gets up. It also usually fills out when he coughs. If left alone it continues increasing in size, so that instead of the bowels being contained, as they should be, in the belly, the greater part drop into the swelling, which may become of an enormous size; and always in proportion to its size is there less disposition in the contained bowels to return back into the belly when the person lies down. Persons who are ruptured often feel a dragging weight about the pit of the stomach, and after great exertion more bowel comes down, and they have an inclination to be sick, which becomes worse till the rupture has, by lying down and gently pressing, either been diminished to its ordinary size, or put back completely into the belly.

Whenever any one finds a swelling at the groin or navel, especially if it have come on suddenly after exertion, he should lose no time in going to the doctor for the purpose of knowing what its true nature is, for if it be a rupture it will require immediate attention; and from want of this many persons have lost their lives. Females often are subject to a rupture, which, from motives of modesty, they conceal, and ask for no assistance till it is too late. The celebrated Queen Caroline, wife of George II., lost her life in this way.

If the medical man determine the swelling to be a rupture, he will recommend a truss or instrument to prevent the bowel slipping out of the belly. If a truss is to be of real use, it should be made specially for the person. The usual way is to go to a truss society or to a surgeon's instrument maker and get a truss fitted.* The difference of these two modes of proceeding may be illustrated by the homely comparison of being measured for a pair of shoes and putting on a pair ready made ; which fits best everybody knows, and just so is it with the two kinds of trusses.

If the person wish to get cured of his rupture, he should never take his truss off, except for a few minutes to cleanse his skin with a sponge, and this even should never be done but whilst he lies down. The truss should always be kept on, whether up or in bed, and if this plan be pursued by young people under eighteen or twenty years of age, and they be careful not to use violent exertion, a rupture will generally be cured after wearing a truss three or four years.

It is always well for those who can afford it to have two trusses of the same size and strength at the same time, so that in case of the spring of the one in use snapping, or it requiring to be mended, the other may be ready to take its place. For if, even for a day, the person be about without his truss, the advantage gained by six or eight months' previous wear will be lost, and all will have to be begun again.

Children whose navel starts may generally be easily managed, by gently pressing the swelling into the navel-hole, and then binding upon it a marble, or a piece of rounded cork sufficiently

---

* Bigg, of Leicester Square and of St. Thomas's Street, Southwark, is one of the best truss-makers.

large to cover, and not go completely into the
hole. The marble will be best fastened with
some strips of strapping put across it, and passing nearly to the spine on each side ; it will require to be removed daily, and the parts having
been sponged clean, then reapplied.

Persons who have ruptures should be especially attentive to keep their bowels regular, so
as to be moved every day, and not allowed to
become costive.

If persons who have ruptures do not wear a
truss, or if the truss they wear do not keep the
bowel up in its place, but allow it to slip down
by the side of the pad, either from the instrument not fitting, or from it not being strong
enough to resist the constant exertions of daily
labour ; or if the truss be worn broken, as is very
frequently the case, the person goes about at
the risk of his life. It may indeed be, that a
person will have a rupture for years, nay, even
for his whole life, may wear no truss, and when
he lies down the rupture may return, or it may
continue down and incapable of being returned
at all, and yet he suffer no inconvenience beyond
a little occasional lassitude. But this is the
exception, not the rule, and from some trifling circumstance or other the bowel becomes strangled,
and the person is certainly destroyed, if not relieved by means presently to be mentioned, or,
on failure of them, by an operation, which, however, is not always sure of saving life.

When a person has been costive two or three
days, and he becomes violently and frequently sick,
at first throwing up stuff like coffee-grounds, and
after some hours, like stools, and very offensive; if
there be a feeling of a cord tied round the midriff, constant feeling of sickness, much uneasiness
and anxiety, there is great reason for supposing
that this has something to do with a rupture.
The inquiry must then be made if the person
have any swelling in the groin or its neighbourhood, at the navel, or anywhere else upon the
belly; and if there be such swelling, how long
it has been, whether it ever returned, or could
be returned, and when it first was unable to be
returned; and whether a truss had been worn or
not; whether the rupture had ever been thus
fixed before, and whether it had been accompanied with vomiting and costiveness, and how
these had been then relieved.

If it be answered that the swelling had always
disappeared on lying down, or that it could be
returned whilst in that posture, till such a time,
when the person had over-exerted himself, the
swelling had suddenly become larger, and would
not return with all his efforts, that then he had
become sick and vomited, and that the vomiting
had become more severe and more frequent, and
that the bowels were costive, and whatever medicine was given, or other fluid taken, was almost
immediately rejected; then there is little doubt
that the bowel has become strangled, and that

its contents cannot pass through it to be voided as stools, and are consequently vomited.

What is then to be done?—It very commonly happens that persons who do not wear a truss, and who are aware of the nature of their complaint, now and then have some difficulty in getting their rupture up, and feel sickish and uncomfortable from it being rather larger and less readily inclined to return than usual. They learn, however, some mode of handling the part, and after more or less effort, replace it, and therefore often rely upon their own capabilities, till they get into a very dangerous condition.

In these cases a medical man should *always* be sent for immediately, for if circumstances should require the performance of an operation, its *safety* mainly depends *on its being performed early*, and its *danger on its being delayed*. Whilst the doctor is being fetched, the patient may be put into a warm bath up to his neck, and kept there till he feel very faint : he may then attempt according to his own usual method to put the rupture up, by pressing it gently, if it be in the groin, or by lifting it up if in the purse, and gently squeezing it towards the belly, but no violence must be used, or the gut will burst.

If this do not succeed, cold may be applied over the swelling, by half filling a bladder with pounded ice and a small handful of salt, or with a freezing mixture consisting of Glauber salt and sal-ammoniac, to which some water must be added. Either of these, after being kept on some hours, will occasionally cause the return of a rupture, but they require to be used with some caution, as if the skin become frosted it may mortify. If neither ice nor the materials for the freezing mixture can be obtained, a wet rag may be put on the part, and evaporation encouraged by a continued stream of air from a pair of bellows, repeatedly wetting the cloth as it dries ; by these means almost as great a degree of cold can be produced as by ice.

Some French surgeons have of late strongly recommended attempting reduction of a rupture by reversing the position of the body, in other words, by putting the patient on his head, or nearly so : and they state that in many instances this method has succeeded. If no medical man can be obtained, it is certainly well worth while to try this plan.

Bleeding a person to faintness often renders the return of a rupture easy, but it should not be used by an unprofessional person, except in situations where no doctor can be had.

It is useless to give purgative medicine by the mouth, as it will certainly be thrown up again, and only renders the vomiting more severe and distressing. The same objection, however, does not apply to a clyster, either of gruel and salt, or gruel and castor oil ; these are occasionally useful in assisting the return of a rupture.

PILES

CONSIST in the enlargement of the veins at the extremity of the lower bowel; and very commonly arise from the straining necessary to relieve the bowels in habitual costiveness. By degrees one or more of the veins become gorged with blood, part of which not being returned, but lodging there, a little tumour is formed, which gradually enlarges. Sometimes it remains within the bowel, and is only forced down at the time the bowels are relieved, after which it either soon returns, or can be pushed up with the finger. But after a time it is continually down, about the size of a small bean, and becoming irritated by the perspiration, it often chafes, and renders walking very distressing. If the bowels be more costive than usual, and their relief need more than ordinary effort, the pile becomes much gorged with blood, which is unable to return. The little swelling then becomes very full and black with the blood it contains; is very painful, often inflames, and either bursts and empties itself, which affords immediate ease, or it runs on to the formation of an abscess, which after a time discharges, and not unfrequently lays the foundation of a fistula. Occasionally the veins inside the bowel become very large and their walls thin, so that the mere effort of relieving the bowels,

without much straining, causes them to bleed,
and sometimes very freely.

Treatment.

The most likely mode of preventing the formation of piles is attention to the bowels, which
ought to be relieved daily once or twice, and
without forcing. Their proper action is equally
beneficial to the health and temper, of which the
celebrated Lord Chesterfield was well aware,
in the advice he gave to a suitor of making
certain inquiries of a great man's valet before
venturing to ask a favour. It is best that nature
should be invited at regular periods. There is
no need of continually taking medicine, if the
person be of a naturally costive habit; for if the
relief be made to depend on this, the quantity
taken will sometimes need to be very considerable, as in proportion to the frequency of taking
purgative medicine does its dose require to be
increased. Regular exercise is generally sufficient to excite the bowels to proper action; but
some persons find it convenient to take a tumbler
of cold water immediately on leaving their bed,
which serves as a gentle and sufficient laxative.
If, however, this be insufficient, it is a most excellent practice to throw up into the bowel, every
morning, half-a-pint or a pint of lukewarm water,
either with an Indian-rubber bottle, or one of

the many syringes for this purpose, which are now so common.

When there is a disposition to piles, the person should always, after relieving the bowels, press up very carefully any little knot or swelling which can be felt, and generally by attention to this, and keeping the bowels gently lax, the ailment is got rid of. It is advisable also to bathe with cold water and to pass up the bowel a small portion of lead or gall ointment; and a teaspoonful of lenitive electuary may be taken occasionally.

When a pile becomes very much swollen, full, and cannot be emptied by gentle pressure continued whilst the patient is lying down, and by the application of linen dipped in cold water, it will be necessary to put on two or three leeches, which may be repeated once or twice; and this will be more especially needful if the pile have become inflamed. Great relief under the latter circumstance is often obtained by freely opening the pile with a lancet, and letting out the clot of blood which often then has formed; this, however, had better be done by the doctor. The person troubled with swollen or inflamed piles should keep at rest for a few days in the horizontal posture, and when the irritation ceases may get up.

CHILDREN who have been much relaxed, from whatever cause, sometimes have a large portion of the bowel protruded, and the same result often happens from nurses putting them on their little chairs, and carelessly allowing them to sit and strain for ten minutes or longer. I have often seen three or four inches of the bowel down from either of these causes. Sometimes it returns with slight pressure, and remains; but at other times is only replaced with great difficulty, as it has become more swollen the longer it is down. At other times, when the protrusion has been of long duration, and of great length, the bowel can be pushed up easily enough, but directly the fingers are removed it protrudes again.

Treatment.

When there is the slightest appearance of protruding gut, the child should not be allowed to sit on his chair for more than two or three minutes; and if the bowels be unrelieved, he should be placed on it again at some little interval. The nurse should carefully notice if any bowel be down, and if it be, should lay the child down on his back with the hips a little raised, and then with her finger gently press it up. If the protrusion will not remain up of itself, a T bandage should be put on, and to the tail-piece half a small bottle cork, rounded at one end and covered with linen, should be fastened and made to rest against the bowel. If the protrusion be very considerable, the bowel must be grasped with all the fingers, emptied as far as possible of the fluids in its substance, and then steadily and gently pressed up. If it be much swollen, this will frequently occupy some time ; but the effort to replace it must be persisted in till the bowel return, otherwise the piece of gut will slough off, and the child may be destroyed. In all cases when much bowel has descended, the child should not be allowed to sit up, but should be kept in the horizontal posture till the bowel have recovered strength to retain its natural place.

WETTING THE BED

Is a troublesome complaint to which children of both sexes, to twelve or sixteen years, are frequently subject; and occasionally, without any apparent cause, this will come on with adults. It often depends on mere idleness and indisposition to get out of a warm bed to empty the bladder. But occasionally it rests on some other cause, not always very apparent.

Now the ordinary mode of treating this affection is to begin with scolding for the presumed misbehaviour, and to end with the hasty parent's or tutor's panacea, as least troublesome to them, for all evil doings, a flogging, often repeated and with great severity, till the part on which the punishment is inflicted and the child's disposition are rendered equally callous, though the practice remain unchecked. There can be no doubt that wetting the bed is with many children an idle trick, or from fear of getting out of bed in the dark, and may be cured by birch reasonably applied. But in numerous instances it is not so, and therefore beating a child, so unhappily circumstanced, from day to day, is downright cruelty, and the inflictor of the punishment, whether parent or tutor, deserves twice as much himself.

Treatment.—Patient watching must be resorted to, that we may ascertain whether this be an idle habit, or an uncontrollable and often unconscious proceeding in the child. If the former, he should be seen to empty his bladder completely immediately before getting into bed, and he should be provided with a little vessel, which, if requisite during the night, he may take into bed, and relieve himself. He may also be threatened punishment if guilty, and should it so happen, correction may be administered with caution, but not with severity, nor repeatedly, since, as already noticed, the child himself may not be really in fault. If the practice be persisted in, either idly or unconsciously, the best mode of treatment I know is to put a blister, about twice as large as a crown-piece, according to the child's size, on the very bottom of the spine, immediately above the cleft of the buttocks. It may be requisite to apply a blister two or three times, at an interval of a fortnight or three weeks, during which it heals and the skin gets sound. In this way I have not very unfrequently succeeded in curing this tiresome ailment.

A little instrument called "a yoke" which may be obtained at a surgical instrument maker's, may also be used without pain or inconvenience, with some benefit; but I prefer a blister.

———◆———

As most people know from experience, is a very painful swelling upon the last joint of the finger, which more or less quickly runs on to the formation of matter. It arises oftentimes without any known cause; at other times, from a hag-nail which has been teased; sometimes from the finger having been bruised, or having been brought to the fire when very cold; not unfrequently from a needle or a splinter running into the finger; and, among laundresses, it is not uncommon from any little crack or wound in the finger becoming irritated by the impure soda or pearlash they use in washing.

There are several kinds of whitlow.

The most common and slightest form occurs generally at one side of the root of the nail, beginning with a little inflammation and throbbing, and by degrees a whitish half-transparent bladder is formed, extending more or less round the nail, the whiteness depending on the thickness and clearness of the skin. If not opened, it continues separating the scarf-skin from the true skin beneath, till it find some crack or thin part in the scarf-skin through which it bursts, discharging the matter, and sometimes there the business ends, new scarf-skin being formed below, and the old bulging, like the skin of a blister, may be picked off. But if the matter have been pent up for a few days, it frequently ulcerates the true skin, and when it has escaped or been let out, a little red body sprouts up through the hole, and, spreading, assumes a cauliflower appearance, the unyielding scarf-skin girting it tightly. This is generally excessively tender, and can scarcely bear touching.

Treatment.

As soon as the blister rises distinctly above the neighbourhood, a pretty large piece should be snipped out, so that the watery matter may readily escape, and continue to flow out as fast as produced. A bread and water poultice should be put on, and in the course of a day or two the cure is completed. But when the little red body of proud flesh, as it is called, has shot forth, it is more troublesome and painful to manage, as all the scarf-skin must be cut off with a pair of scissors, to free it entirely from the narrow hole. The pain of this removal must be borne, for otherwise the proud flesh will continue increasing, and become more and more tender the longer it is delayed. Generally after poulticing a few days, and binding up lightly with some mild ointment, this sore also gets well. Not unfrequently, in both cases, if the whitlow have run completely round the root of the nail, the nail is destroyed, and slowly pushed off as another is formed at the quick or root, and continues to grow.

The *second kind* of whitlow is formed in the bulbous front of the end of the finger; this is much more severe, as the matter is formed deeper, and the scarf-skin being tougher, will not give way to the matter which desires to escape. The pain extends sometimes into the hand and arm, and the matter occasionally burrows up the finger.

Treatment.

Cut through the skin freely and deeply with a knife, in the direction of the length of the joint, and put on a poultice, which will afford almost immediate relief from agony.

This, when very severe, will occasionally run into

The *third kind* of whitlow, which is worst of all, and may also occur independent of the former. In this case, the sheath which contains the tendons of the finger inflames, the finger swells, and unless quickly attended to, the inflammation spreads into the hand, and the tendons and one or more of the bones of the finger are destroyed; or at best the finger becomes shrivelled and stiff.

Treatment.

Leeches here should be frequently applied;

warm bathing of the whole hand, which should
be wrapped up in a bread and water poultice. If the doctor can be had, let him make a
free cut down to the bone, at latest, after twenty
or thirty hours have elapsed from the beginning
of the attack. It will give greater relief from
the severe pain, and be more likely to check the
mischief at once, than if the cut be delayed,
which must be made at last.

BOILS,

As most people are aware, are great plagues, not
only from the severe pain which accompanies
them, but also from their frequent recurrence for
weeks or even months after having once made
their appearance. They are generally in themselves of no great consequence, except when from
their great size the setfast, core, or slough which
is formed by them is very large, and its separation
causes much constitutional excitement, as in a
carbuncle, which is only a very large boil ; or when
from their situation near the opening of the lower
bowel they produce, as they do commonly at that
part—a Fistula.

Treatment.—Bathing with warm water and
poulticing should be employed, and a cut with a
knife or lancet through the inflamed skin which

covers the core should be early resorted to, as it relieves the pain by getting rid of the pressure of the unyielding skin, and allowing the escape of any fluid collected beneath, whilst at the same time it hastens the separation of the core.

As boils and carbuncles especially show signs of weakened constitution, the health should be improved and generous diet allowed.

BLACKHEADS,

As they are called, are very frequent on the forehead and by the sides of the nose of unhealthy young people. These are merely the blackened ends of a greasy matter which is constantly produced more or less largely over the whole surface of the body, to defend it from the pungency of the perspiration, which would otherwise make the surface raw, as frequently happens in chafing, when the perspiration collects in quantity, where any one part of the body laps immediately against another, as for instance in the armpits. This greasy matter sometimes, instead of oozing from the little pits in which it is formed, becomes thickened, cannot of itself escape from its little tube, and has its top blackened by the dirt. But it may be squeezed out by a little gentle pressure in shape of a small delicate thread, and from its resemblance to a little white worm is commonly called a worm or maggot.

Treatment.—Bathing the face with warm water and rubbing briskly with a hand towel is usually sufficient to dislodge these ugly ornaments to the face; or if this be not sufficient, a little gentle pressure with a finger on each side of the black-head, will often squeeze them out. But violent squeezing must not be used, as it is very likely to lead to the formation of little painful boils, which will run on for three or four days before the core can be pressed out. Not unfrequently indeed, when left alone, these little collections of grease irritate, their tubes inflame, matter is formed around them, and very ugly pimples appear on the face. When this happens it is better not to meddle with them further than to bathe with warm water, till matter is seen to have formed, when a little conical pustule, with a white head dotted with black, shows it in a fitting state to prick with the point of a lancet or a sharp needle; then pressure applied as mentioned already, easily empties the whole, and the pimple soon heals.

INGROWING NAIL.

ONE of the deserved punishments which people suffer for the folly of squeezing their feet into narrow shoes and boots, is an ingrowing nail. The toe usually attacked is the great one, and by the continued pressure on its sides, the nail, instead of its ordinary form, becomes narrow, and much arched across. In this state the pressure of the shoe forces the corner of the nail down into the skin. At first there is only a little uneasiness, which ceases when the foot escapes into a slipper. By-and-bye the pressed part feels a little sore, a slight moisture is found on the stocking, and on examining the toe, the pressed edge is found inflamed and tender. Perhaps the person discovers that the nail digs in, and he endeavours to cut away the pressing part; but this usually only makes matters worse, by leaving a sharp corner, which is driven in more deeply as the nail becomes more arched by perseverance in wearing the tight shoe. The sore now attempts to relieve itself of pressure by producing a little mass of proud flesh, which, however, only adds to the mischief, and at last the person becomes so lame that he is almost unable to move.

Treatment.

First get rid of the narrow shoe, so that the toe may be unconfined, and the nail allowed to recover its proper breadth, which, however, it does not do very quickly. Then proceed to relieve the sore skin by the side of the nail of its pressure. It is of no use, however, merely to cut away the pressing nail even freely, and then to thrust a piece of lint under its edge, which is as painful as it is useless; for the nail, if not otherwise managed, will drop, in the course of a few days, upon the old spot, and again render it angry. The proper treatment is, thinning the whole length of the middle of the nail, from its root to its end as much as possible; and this is best done by scraping it perseveringly with the sharp edge of a piece of glass again and again, till the middle of the nail be as thin as writing-paper, and will readily bend under pressure of the finger-nail. This is at first a rather painful job; but the scraping must be done with a light hand. As soon as the middle of the nail has been thus thinned, it yields to the upward pressure of the skin on its side edges, readily bends, and offers no further resistance. The sore place being then no longer irritated by pressure, the proud flesh soon drops down, and the sore heals. Some persons recommend inserting a bit of lint between the nail and the proud flesh, but I do not, for it is painful and troublesome to push in, and is unnecessary.

If narrow shoes or boots be again used, the foolish wearer may expect a repetition of his plague.

BUNIONS.

Young ladies and gentlemen, for the sake of making their feet small and pretty, as they fancy, are frequently in the habit of wearing boots and shoes which are too small, and, as a just punishment for their folly, cripple themselves and spoil the shape of their feet, by the pressure acting especially either on the ball of the great toe or on the instep, and producing bunions. The beginning of a bunion is first shown by pain and redness, with a little swelling on the part of the foot particularly pressed. The pain at first soon ceases after the boot or shoe has been taken off; and if it be not worn again for a few days, all the angry feeling subsides. But if the shoe be still worn, the pressed part, which is still tender, becomes more painful, swells more, and the redness will sometimes spread to the size of half a crown; and the shoe cannot be worn without slashing it over the pressed part. The swelling, if the pressure be not continued, loses its redness and tenderness, but feels as if full of fluid, and after a time becomes hard and solid. If at the ball of the toe, it juts inwards, and completely spoils the shape of the foot, and nothing will recover it. If on the instep, it does not so much matter as to appearance, for it only makes a very high instep ; but it is no less painful than at the toe, and increases in proportion to the continuance of the pressure. Sometimes these bunions, specially at the toe, inflame so considerably that they go on to the formation of matter, and very troublesome lameness is produced.

Treatment.

This consists in removing all pressure from the part. The formation of a bunion may in the beginning be prevented, but only in the beginning ; for when once actually formed, it is scarcely possible ever to get rid of it, and it remains an everlasting plague. To prevent the formation of a bunion, it is necessary, whenever and wherever a shoe or boot pinches, to have it eased at once, and so long as that part of the foot pinched remains tender, not to put on the offending shoe again. When a bunion has once completely formed, if the person wish to have any peace, and not have it increase, he must have a last made to fit his foot, and have his shoe made upon it. And whenever the bunion inflames and is painful, it must be bathed with warm water, and poulticed at night.

LIKE bunions, depend on unequal pressure on
some part of the foot; generally upon the upper
surface of the smaller toes, but sometimes between them, and occasionally also upon the sole
of the foot. But they differ from bunions in
being formed only of an increased production of
the scarf-skin or cuticle. Upon the pressed part
the skin becomes tough to some extent, and
presses upon the sensible skin beneath, which
endeavouring to relieve itself from the pressure,
produces an additional quantity of scarf-skin at
the part pressed, for the purpose of pushing it
off. So long as the pressure continues, this formation of new scarf-skin continues, but to less
extent, so that the corn now formed has a cone-
like shape, of which the base is on the surface of
the toe, and the point below pressing into the
true skin. A more homely and not unfitting
.ikeness may be found in a short clump-nail, the
top of the nail answering to the surface of the
corn, and the point to the point· of the corn.
Either of these comparisons readily explains why
a corn is rendered more painful by a tight shoe ;
for its top being pressed on, necessarily thrusts

the point into the sensible skin, and if the pressure be continued, it sometimes causes so much
inflammation, that an abscess forms at the root
of the corn, as its descending part is called,
which produces excessive suffering.

Treatment.

It is of no use merely to cut away the hard
skin that forms the top of the corn, nor even to
pick out the root with the finger-nail, or with
the point of a pair of scissors, if the person will
be foolish enough to persist in squeezing his feet
into shoes that are too tight, and make undue
pressure. The treatment consists in relieving
the part where the corn is situated of all pressure
and as far as possible preventing any continuance
of pressure. The foot should therefore be well
soaked in warm water, which softens the corn;
then all the thick skin on the surface should be
carefully picked off with the finger-nail, or removed with scissors or a penknife, taking great
care not to wound the sensible skin beneath, as
if this be made to bleed, it often inflames, and
the toe becomes very painful for a few days.
The broad spreading top of the corn having now
been removed, a thin, soft layer of scarf-skin
only remains, except in the very centre, where
a little white, hard, almost horny substance, is
found, which is the point or root of the corn, and
if pressed down causes smart pain, by driving

against the sensible skin beneath. Now, as much of this as possible is to be dug out with the point of the scissors, so long as it can be done without giving pain and making the part bleed; for when the digging becomes painful, it is a sure sign that the sensible skin is very close by. If the root of the corn be nearly and pretty well got out, a little conical cavity is left in which it had been lodged, and this is the part which has to be well protected against pressure, at least till the true skin below has ceased to be irritable. For this purpose, it is necessary to have some thick buff-leather, spread on one side with soap-plaster, rings of which must be cut out and laid one above the other, the first having a hole sufficiently large to give the corn " a wide berth," as the sailors call it—that is, plenty of room; and the hole of each succeeding ring being less than the former, till the topmost has no hole in it. In this way of placing one plaster above the other, a conical cavity is also formed, so that the seat of the corn is entirely freed from pressure. And from suffering extreme pain, the person is often at once put in ease and comfort, and can walk miles, though previously he could scarcely hobble across the room. The number of layers of leather will depend on its thickness, but generally three or four are sufficient.

If inflammation have taken place at the root of the corn, or if the true skin have been wounded in attempting its removal, and inflammation follow, then it will be necessary to soak the foot well in warm water, put a poultice on the toe, and keep at rest with the foot on a sofa, otherwise the consequences may be serious, and even fatal.

Corns now and then form on the palms of the hands, specially in persons unaccustomed to laborious occupation, who get their hands much rubbed, without being blistered, in rowing, or handling a hammer or a turning tool. This kind of corn is rather unsightly than painful, and consists only of a wide-spread thickening of the scarf-skin without any root. They are best left alone, and after the unwonted hand-labour has been left off they soon disappear of themselves.

A STYE IN THE EYE.

UNHEALTHY children are very liable to the formation of little abscesses between the roots of the eyelashes, and which are commonly called sties. Sometimes one or two form at the same time on the edge of one or both eyelids ; and are continually being formed at intervals of a few days, sometimes for weeks. They rarely become larger than a small pea ; at first are hard and red, smart and ache, and in course of two or three days matter forms in them. They either burst by the mere motions of the eyelids, or, as they render the part uneasy, the child rubs them till they break, and then a little scab forms, which soon drops off.

Treatment.

They are best managed by bathing frequently with warm water, or with warm poppy-water if very painful. When they have burst, and the scab falls, an ointment, composed of one part of citron ointment and four of spermaceti ointment, well rubbed and mixed together with a bone paper-knife, should be smeared along the edge of the eyelid every night at going to bed. A grain or two of calomel, with five or eight grains of rhubarb, according to the child's age, should be given twice a week. By these means the complaint may be checked or got rid of ; but it is often very tiresome and unmanageable.

A BLIGHT IN THE EYE.

It is not unusual when a person receives a blow on the eye, that some vessel bursts and blood is poured out between the outermost coats of the eyeball, so that the part usually white becomes red or almost black; this is in reality nothing more than a bruise.

But occasionally it happens that when on going to bed nothing uncommon is noticed on the eyeball, yet on the following morning when standing before the glass the person is frightened by finding the white of his eye has turned dark red or nearly black. I believe this is commonly known as "a Blight in the Eye." It has a very ugly appearance, but is unaccompanied with pain or any inconvenience, and excepting its look, is of no consequence, as it generally subsides in a week or ten days without any treatment. How it arises is by no means clear; probably a little vessel has from some accidental cause been distended beyond its strength, and bursting, the blood escapes and quickly spreads itself between the outer coats of the eyeball.

s 2

TUMOURS IN THE EYELIDS.

Not unfrequently people are plagued with a succession of little tumours, as large as a small pea, in the eyelid itself, usually in the upper lid. Generally only one of them is formed at a time, and is very distinctly seen beneath the skin. After a time they commonly make their way through the little gristle which stiffens the eyelid, and bursting, their size gradually diminishes, till at last they either empty and refill again, or disappear entirely.

Treatment.

Sometimes it is recommended to cut them out, but unless they become very large it is not needful; and frequently, indeed, they are formed again so often that people are tired of having them removed. After a time they usually cease, without any seeming reason, to be produced. On the whole I believe it is better not to take them out; and merely so soon as they begin to appear, to slip a little citron ointment, diluted as for sties, under the eyelid every night with the end of a feather. For a few minutes its application produces a slight smarting, but this soon goes off, and perhaps in course of ten days or a fortnight the little tumour disappears.

INFLAMMATION ON THE SURFACE OF THE EYE.

SLIGHT inflammation of the membrane covering the globe of the eye, and lining the insides of the eyelids, is not unfrequently occurring when the eye has been exposed to a current of cold air. The eye waters, feels as if sand were in it, the white part is reddened, and soon after a little matter is formed, and on waking in the morning the eyelids are felt glued together. Such is the common condition of a slight degree of ophthalmia, as this inflammation is called, which, however, may run on and become so severe, that the eye may be destroyed by it very speedily. Or if the attacks recur frequently, the transparent part of the eye may become so completely dulled that the sight is lost.

Treatment.

When in the mild form I have mentioned, it may usually be got rid of by one or two smart purges with calomel and rhubarb, and bathing the eye with warm poppy-water. If after all the pain and redness have ceased the eye feel weak, it may be improved by washing it frequently during the day with a lotion composed of a grain of sugar of lead to a large tablespoonful of soft water.

Infants of a few days or weeks old are very liable to severe attacks of inflammation of both eyes; the lids quickly inflame and swell, and large quantities of matter are discharged from between them. This is a very dangerous complaint, as in the course of two or three days one or both eyes are very frequently destroyed. It is always advisable directly these symptoms make their appearance to send for the doctor at once, as no time should be lost. If he cannot come immediately, it is right to clean out this matter several times a day, by carefully squirting some warm water, with a small bone or pewter squirt, between the eyelids. And if no medical aid can be obtained, on the second day a solution of alum, in proportion of a grain or two to two large table-spoonfuls of soft water, should, after first washing out the eye as already mentioned, be thrown between the lids with a squirt. This kind of inflammation is always much to be dreaded.

The person who syringes the child's eye should *be careful not to get* any of the matter into his own, as the same disease will be produced and speedily destroy the eye, often even under the best treatment.

PUSTULES ON THE EYE.

CHILDREN not unfrequently have form upon the front of the eyeball, either on the transparent or on the white part, a little pimple rather bigger than a millet-seed, sometimes accompanied with inflammation, and at other times without any. Although when the little pustule breaks it generally leaves a little ulcer upon the white of the eye, it is not usually of great consequence; yet if upon the transparent part, it is serious, as when the little sore heals, it often leaves a small scar, or speck, as it is called, which interferes with the sight.

Treatment.

Not unfrequently these little pustules disappear after one or two doses of a grain or two of calomel and a few grains of rhubarb, with a small blister on the temple, and bathing the eye with warm water or poppy-water. But if the pustule go on to bursting, and a little ulcer form, after two or three days, if there be no surrounding redness, the ulcer should be touched with the point of a camel's-hair pencil dipped in a solution of lunar caustic, in the proportion of one grain to an ounce, or two large table-spoonfuls, of soft water.

As, however, the eye is as valuable as it is a delicate organ, it is always best to obtain medical aid, if it can possibly be procured, and as soon after it has been attacked by disease as can be managed; for the delay of less than twelve hours may make the difference between saving an eye, and irreparable blindness.

MILK ABSCESSES.

MANY females, within a few weeks, or sometimes even days, after their confinement, are attacked with inflammation of the breast, in consequence of the milk then continually produced not being emptied. This may depend either on the mother, forgetful of her duty, and too anxious about her own person and of her ease, refusing to suckle her child ; or from inflammation brought on by an accidental chill closing the little vessels by which the milk naturally passes from the nipple. In either case the effect is the same, the breast slowly becomes gorged with milk, and at last active inflammation is set up, which runs on to abscess, and one or more openings are formed through the skin, by which the pent-up milk, mixed with matter, is discharged.

The breast at first becomes full and uneasy ; then hard and painful, and the pain is soon sharp and shooting, becoming more severe every time the draught comes into the breast. The skin inflames more or less widely, but at last generally the inflammation becomes less in extent, but more severe, and attended with great throbbing at one particular part. The skin now changes colour, protrudes, has a shining hue, and, becoming deep

red or blackish, gives way, and the contents of
the abscess which has been forming are discharged,
and the patient is relieved from the severe pain,
but a deep ugly sore is left, which generally
heals very slowly, and leaves a very unsightly
scar.

Treatment.—At the onset of this complaint,
the greatest effort should be made to empty the
breast, by the natural means of suckling, or by
drawing the breast with a proper milking tube ;
and if the milk can be in this way emptied, the
fulness and uneasiness usually subside.   But if
these means fail, it will be proper to endeavour
to check the flow of milk to the breast by applying
cold evaporating lotions, and giving Epsom salts
in sufficient quantity, a large teaspoonful three
or four times a day, to act very freely on the
bowels, which sometimes answers the purpose and
prevents further progress.   Should, however, the
inflammation remain unchecked by these remedies, and the swelling and redness increase, then
leeches—ten or a dozen—may be applied, and
repeated two or three times, followed up with
bread and water poulticing.   If matter have
once formed, and when it has, it is generally
accompanied with one or more fits of shivering,
little, if anything, can prevent it making way to
the skin ; and the object then is, to prevent the
ugly scar produced by the destruction of the skin
over the abscess.   For this reason it is best, as

soon as the fluid can be felt, even though deeply, beneath the skin, to have an opening made with a lancet, which, if possible, should be done by a medical man. The pain of this is indeed very severe, but it is momentary, and the relief is great within a few minutes, as the milk and matter escape immediately by the opening, and continue so doing till the wound fills up from the bottom. Those, however, who have not courage to have the abscess punctured, must bear their agony till it break, and with the almost certain prospect of a scarred breast. Generally it is only necessary to continue the application of the poultice, and to keep the bowels constantly acting with Epsom salts till the milk ceases to form ; after which, in the course of a fortnight or three weeks, the wound usually heals.

ARE very frequently a source of great annoyance during suckling, and not rarely is the agony so great when the nipple is pressed by the child's lips, that the fondest mother dreads every time she takes the child to her breast. The nipple is also often sore and raw, and discharges an acrid fluid which increases the soreness.

Treatment.—This is often very difficult, for though many applications might be made which would quickly heal the soreness, yet their taste would be so disagreeable to the child that he would not take the breast. It is better, therefore, to trust specially to frequent bathing with warm or cold water, as most agreeable to the patient's feelings, so as to prevent any collection of the acrid fluid and increase the soreness, and to dab the nipple frequently with a little cold brandy and water, or spirits of wine and water, which dull the tenderness of the part, and do not by their taste indispose the child to suck. It is also advisable that the nipple should be protected by a shield of silver or ivory, the former being best, as it can be kept clean, covered with a cow's teat, which is easily obtained, and may be kept in a little weak spirit and water. The shield should always be sufficiently large to receive the nipple without squeezing, and to allow its swelling as the milk is drawn through it ; and it should only be worn whilst the child is being suckled.

IRRITABLE BREAST

YOUNG unmarried women are often sadly troubled with tenderness and pain, and even with lumps in the breast, becoming worse at particular times. This condition not unfrequently continues for months, and is not only very harassing from the severity of the pain, but from the fear which is had of it being a cancer. Further than the annoyance, it is rarely of any consequence. It is a frequent ailment of that time of life, and often attacks first one breast and then the other, or even both at once.

It is best to have a medical opinion when the breast continues painful for a length of time, and more especially if there be any lump in it. But if it be such as here mentioned—and I have merely hinted at it on account of the frequent dread of its dangerous character, there is really no cause for alarm

BEFORE proceeding to consider the various ways by which Stifling may be produced, and the means of restoring persons who are seemingly dead from this cause, it will not be out of place to give a general and simple account of the mode in which Breathing, or Respiration, as it is technically called, is performed, so that it may be understood what is the immediate cause of stifling, which cannot be otherwise explained.

The Bible assertion in regard to animals, that "in the blood is the life thereof," is indisputable, and will remain so, spite of the sneers of those too knowing to bow to Revelation, and of the cavillers with JOHN HUNTER, whose extensive physiological inquiries (greater than those of any other man, as his Museum and his writings prove) made him a stanch supporter of this great doctrine. By this important fluid, which is distributed through every part of the living being, the body is not only sustained, but the several parts, flesh, bone, and other organs, of which it is composed, are built up, till the animal has attained its full bulk. And as these several parts decay and become hurtful, the blood is the great stream which, receiving all these decayed parts that have become impurities, and are hurtful to the body, carries them to particular organs, which are destined to get rid of them ; the most remarkable examples of which are exhibited by the production of urine in the kidneys, of the perspiration by the skin, and the discharge of carbonic acid gas by the lungs. With the last we have only to do.

The circulation of the blood, and respiration or breathing, have a mutual relation to each other. The circulation or carrying of the blood is performed by the heart and its vessels ; the respiration or breathing, by which the blood undergoes peculiar changes, is effected in the lungs.

The heart consists of a pair of fleshy force-pumps, called ventricles, by which the blood is driven through the arteries, which are the conduit pipes *from* the heart, by the right ventricle into the lungs, and by the left into all parts of the body ; both ventricles act at the same time. Each ventricle is provided with a reservoir, or auricle, to receive the blood from the veins, which are conduit pipes *to* the heart. The right auricle receives all the blood, excepting from the lungs, from all parts of the body, and pours it into the right ventricle ; the left receives the blood from the lungs alone, and pours it into the left ventricle.

The lungs consist of two large bags, divided, by innumerable small partitions, into very mi-

nute cells, of less size than the smallest pin's
head, each of which can receive air by means of
correspondently minute tubes, or extreme branch-
ings of the windpipe, which collecting together
form larger branches, and these again uniting
form still larger, and continue uniting again and
again till each lung have only one great air-tube,
and then these two joining form the windpipe,
which ascends up the neck before the gullet, and
has attached at its top the organ of voice, or
glottis of anatomists, through which the windpipe opens by a very narrow slit or chink (*a*) into
the throat, in front of the entrance into the gullet.
Now, as the food, in passing from the mouth into
the gullet, must pass over the chink of the windpipe, we should be liable to the great danger of
it slipping into the windpipe; nay, even every
time we swallowed our spittle we should run
like risk, were it not that this chink is covered
and protected by a very simple, but beautiful
contrivance, which falls down upon it in the act
of swallowing, and forms a shoot, over which
solid or fluid slips, without danger, into the top of
the gullet.   This little shoot is called the epiglottis, because it is above the glottis, or voice
organ.   Whilst we are breathing, the epiglottis (*b*)
stands upright, as seen in the sketch A, and the
air passes readily in and out of the windpipe.
When we swallow, the epiglottis (*b*) falls upon the
chink of the windpipe, as in sketch B, and shuts

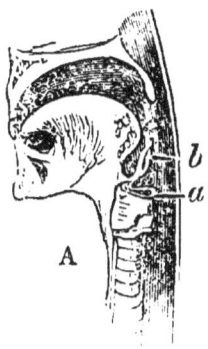 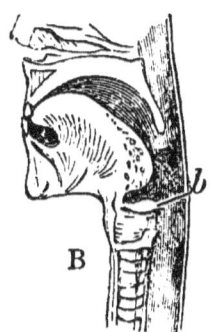

it closely, so that neither can the air pass in
or out, nor the solid or fluid food get in. But
it must be remembered that the epiglottis is
neither pushed down by the food as it passes
to the gullet, nor pulled down by any other contrivance. On the contrary, it is the top of the
gullet, which pulled up to receive the food as it
passes into the throat, lifts up with it the voice
organ, and pressing it against the bottom of the
epiglottis, the loose end of the latter drops on it.
Having thus given a brief description of the apparatus by which circulation and respiration are
performed, let us now take a slight view of the
process of respiration and its use.

It has been mentioned that the blood having
performed the office of building up and sustaining the body by communicating some of its living
properties becomes spoiled, and when it has
been received from the minute utmost branches
of the arteries, into the corresponding minute

T

branches of the veins, is carried on by the larger
trunks of the veins to the heart, collecting, as it
runs along, more and more blood, which contains
in it the worn-out and useless parts of the body ;
and thus when it reaches the right auricle or reservoir of the heart, it has become utterly spoiled
and unfit to support life. Now comes the need
for breathing ; this spoiled blood contains a quantity of carbon or charcoal, which is the cause of
its dark colour in the veins, and this can only be
got rid of by exposure of the spoiled dark-coloured
blood to the action of atmospheric air, which is
done in the lungs. The atmospheric air consists
of nitrogen and oxygen, and being drawn in by
inspiration, or drawing in the breath, as it is vulgarly called, descends through the windpipe and
its branches, into the most minute cells of the
lungs, in the walls of which branch the termina-
tions of the pulmonary artery, or great artery
from the right ventricle, and the beginnings of the
pulmonary veins, which communicate with the
left auricle.

Now the process of respiration may be thus
shortly explained. The right auricle being filled
with the spoiled blood brought to the heart, a
disposition to get rid of it takes place, and the
person draws in his breath, or inspires, and all the
air-cells are filled with fresh air. The right
auricle empties the blood into the right ventricle,
which directly forces it into the pulmonary artery,

and by its branches into the lungs, where, upon
the walls of the air-cells, it is exposed to the action of the air, a part of the oxygen from which
unites with the carbon in the blood, forming carbonic acid gas, and by some peculiar change the
blood is again fitted for the support and building
up of the body, and for preserving the animal
heat. The blood is now found of a bright red
colour in the pulmonary veins, and through them
flows into the left auricle. As soon as this change
has taken place, the air, fouled by having lost
part of its oxygen, which has been changed into
carbonic acid gas, is driven out of the lungs and
through the windpipe, in the contrary course to
that by which it had entered, and thus is said to
be expired. The purified blood in the left auricle
proceeds into the left ventricle, is distributed by
it through all the arteries of the body, and is
again returned spoiled to the right side of the
heart, whence it has to pass through the lungs, to
be there repurified, brought again back to the left
side of the heart, and thence again distributed.

Such is the way in which breathing is performed, and such the mode by which the spoiled
blood is purified, or freed from its charcoal during
its passage through the lungs. It must be observed, however, that the two sides of the heart do
not act one after the other; but the two auricles
act together, and the two ventricles act together;
and when the two auricles are expanded by the

blood they have received, the two ventricles are contracted and empty, but so soon as the latter are empty they relax, the blood is forced gently from the auricles into them, and so soon as filled, the ventricles are excited to contract and force the blood from them, which is prevented returning into the auricles by an apparatus of valves, the right ventricle throwing its contents into the pulmonary artery, which takes it to the lungs, and the left into the aorta, or great trunk artery, which by its branches conveys it to all parts of the body.

So long as the blood continues to be properly purified, or aerated, as it is called, in the lungs, everything goes on well; but if the air breathed be impure, so that the change cannot be duly effected, then uneasiness and oppression about the chest increases to distress, is followed by headache, and may end in fainting. There are few persons, I dare say, who have not suffered this inconvenience more or less severely in hot crowded rooms. In these cases the foul air actually received into the lungs is the cause of the uneasiness

STIFLING

DEPENDS on the blood not undergoing that change by which it gets rid of the charcoal or carbon which it has acquired during its return from the extreme branches of the veins to the heart. This condition may depend either on foul air of various kinds being breathed for a certain time, or on the atmospheric air being prevented passing into the lungs, either by the closure of the epiglottis on the chink of the voice organ, or by the windpipe being squeezed so tightly below, that the air cannot descend along it into the lungs. In both cases the result as regards the blood is the same, it becomes less and less freed from the carbon, becomes blacker and blacker, circulates more and more slowly, at last will not flow from the lungs into the left side of the heart, which is consequently found empty after death, whilst from the same cause the right side of the heart is gorged with black blood. Although the effects produced by breathing foul air, and by not breathing at all, are really the same in the end, yet those from the former are in general produced less rapidly, and attempts for their relief are therefore more likely to succeed than in the latter, in which life is destroyed usually in the course of a very few

minutes.   There are other circumstances as
regards the state of the brain consequent on
stifling, which might be mentioned, but are not
called for here.

CARBONIC ACID GAS

Is the air produced in fermentation and in
slaking lime ; it is also found in wells, cellars, or
caves, which have been long closed up.   If un-
diluted with common air it cannot be breathed,
as the moment it reaches the top of the windpipe the epiglottis shuts down close, and none of
it can enter.   Any one may easily convince himself of this, by cautiously passing his head into a
brewer's square where fermentation is going on,
and immediately his mouth and nostrils are
brought within range of the gas, he feels his
breathing checked, which depends on the sudden
shutting down of the epiglottis, and he quickly
draws back from the pestilent air.

Persons are commonly subject to the influence
of this fatal gas, in descending into brewers'
squares, or into wells before the gas has been
sufficiently emptied, by dashing water about the
square, or in the well, till a lighted candle let
down to the bottom will burn freely.   If this be
neglected, and the person descend a ladder into
the square or well, so soon as he comes within
the range of the gas, which is always strongest
near the bottom, down he drops as if shot, and

whoever follows him drops in like manner, so
that two or three, or more persons, as long as
there be any venturesome enough to go down,
will all be lying at the bottom in the same
helpless and dangerous state, and be soon destroyed, unless they can be quickly fished out,
which is a difficult matter. In a brewhouse,
especially in a large concern, every fermenting
square ought to be furnished with a light crane,
with a block and rope having a grappling-iron,
with which the whole bottom of the square may
be swept, so as to fish up any labourer who drops,
in attempting to descend to clean it, before the
air has been properly purified. The fixing of
this tackle ought to be made compulsory by law,
for labourers are often meeting with this accident, in consequence of their perverseness in descending into the square before it is fit for them
to do so, notwithstanding the strongest warnings
given of the danger they incur.

  If no such tackle be ready, a hop-bag hook,
or a pitchfork with its prongs bent up to form a
hook, with a strong cord or rope attached to it,
must be cautiously thrown beyond the fallen
person, and carefully drawn for the purpose of
catching the clothes. A second or even a third
of these hooks may in like manner be lowered,
till sufficient means and power are provided to
hoist up safely. If neither hook nor pitchfork
can be laid, even a rope with a running noose

and a long pole to get on an arm or a leg will be a very good substitute. If the person have fallen into a well, and there be no winch nor rope to fasten a pitchfork to, the means just mentioned must be employed.

The quickest way of getting a person out, if the well be not very deep, and his situation can be seen from above, is by some bold, clear-headed person descending quickly, and seizing him as speedily as possible. But as this is matter of considerable danger, the person who descends should have his head and shoulders well wrapped in a thick sack, through which the foul air cannot easily enter, and a rope securely tied round him and held by the bystanders ready to pull him up, if, in his praiseworthy efforts, himself should be overcome by the foul air.

Treatment.

When the person has been got out, it is best to bring him immediately into the open air, and there keep him, lying on a shutter or board whilst the means for his restoration are being employed. If he have not fallen into water, his body will generally be found heated ; he must then be quickly stripped, and his head, neck, and chest freely dashed with cold water. Or if a brook were close by, his body might be plunged in again and again, taking care, however, not to dip his head in also. This mode of

restoring an animal, which has been half-killed
by plunging into foul air, is practised on the dogs
that have been subjected to the foul air of the
Grotto del Cani, near Naples, and their recovery is much hastened thereby. If, however, the
person have fallen into water, and be cold, then
he must be put to bed in an airy room and warmth
applied by hot bottles to the feet and stomach.
Attempts should be made to empty the lungs of
the air they contain, and to fill them with fresh
air. This may be easily done by one person
pressing, with both his hands, the breast-bone
firmly down towards the back, whilst another,
with his hands outspread, presses as nearly as
possible the whole surface of the belly, which
forces the bowels against the diaphragm, or
great muscular partition between the chest and
belly, and thrusts it up into the chest: by these
means the lungs are brought into almost as
small a space as when a person by his own will
expires forcibly and throws out a large quantity
of air, which lifts up the epiglottis, if still closed
upon the chink of the windpipe. The hands are
then to be suddenly withdrawn, when the breast-
bone, freed from pressure, rises, and increases
the capacity of the chest, which is further enlarged
by the bowels returning to their usual place and
ceasing to force the diaphragm into the chest;
immediately the air rushes down the windpipe
and fills the lungs. This proceeding which must

be repeated again and again, is as efficient a
mode of producing artificial respiration as can be
devised, and far better than attempting to fill the
lungs with a pair of bellows, the nozzle of which
is put into the mouth, and the lips gathered round
it, whilst the nostrils are closed by grasping with
the fingers, for in this way the stomach is more
commonly distended than the lungs. The chest
and limbs may also be well rubbed with some stimulating liniment. Strong smelling salts may be
held to the nose, and the throat tickled with a
feather to excite vomiting. It has been advised
to pass a pipe into the chink of the windpipe, and
fill the lungs through it with common air, or with
oxygen gas. I do not think any great advantage
is obtained by the pipe, and, indeed, it cannot be
managed by unpractised hands, nor do those, who
are capable of using it, do much more good ; and
as regards the oxygen gas, we are rarely able to
procure it, or to make it, immediately. The
same applies to electricity, which has also been
recommended.

AIR POISONED BY BURNING CHARCOAL.

The accident just mentioned depends upon the
action of pure or nearly pure carbonic acid upon
the epiglottis, in consequence of which no air of
any kind can get into the lungs. But sometimes
the air of a room is poisoned by a comparatively
small quantity of carbonic acid gas, which is pro-

duced by burning charcoal in a room, as generally employed to hasten the drying of stuccoed walls. Under these circumstances the foul air is actually received into the lungs, and its fatal effects are felt sooner or later according to the quantity of the gas produced, which is always larger when the charcoal burns slowly, and smaller when it burns quickly. The person, although active and employed at his work, quickly drops, loses all power of moving, or even calling out for assistance, falls into a state of complete stupidity, and dies in nearly the same state as a person who is destroyed by opium.

Treatment.—The same as in the-former case.

AIR POISONED BY BURNING COALS.

This happens when common coal is burnt in a room which has no vent or chimney. Carbonic acid gas is here also an active ingredient; but another very acrid gas is produced at the same time, which, directly it reaches the top of the windpipe, produces a stifling sensation, and the person rushes from the room as quickly as possible, coughing violently. If, however, he have dropped asleep whilst the air was only slightly fouled, it is not improbable that he will be found senseless or dead soon after, more or less quickly, according to the quantity of deleterious air produced.

Treatment.—Same.

The carbonic acid gas produced by burning

lime or bricks produces the same symptoms, and must be treated in the same way.

FOUL AIR IN DRAINS AND PRIVIES

Is rendered so by the production of sulphuretted hydrogen gas from the soil they contain ; and unless some caution be observed in opening them, if they have been long closed, directly the person employed breathes a small quantity of this air, which escapes quickly as he strikes his pickaxe through the vault, down he drops as if dead. But if this air be mixed with much common air, he may breathe it for a little while ; his breathing, however, becomes difficult, and then he loses his strength, falls, and becomes insensible and cold, his lips and face blue, and his mouth covered with bloody mucus. The limbs are generally relaxed, but are sometimes convulsed. In a much smaller quantity this gas produces disposition to sickness, or actual vomiting, much rolling about of wind in the bowels, with inclination to relieve the bowels, and sometimes a quick purging and head-ache.

Treatment.

The person should be removed as quickly as possible into fresh air, beyond the influence of the foul air. His breathing should be encouraged in the way already directed, and if he be very cold, some brandy or other stimulant must be got into his stomach as speedily as possible.

As a means of guarding against the dangerous consequences which may result from opening long-closed drains or privies, it is always advisable, if possible, to convey into them a pretty large quantity of liquor of chlorinate of lime ; or a few pounds of chlorinate of lime mixed up in a pail of water, which is rather better, as the gas produced by the mixture is given off more slowly, and acts more effectually on the foul air.

DROWNING

Is a very common accident, and upon the immediate and proper assistance which the person receives who has fallen into the water and thereby been nearly stifled, depends the probability of his recovery.

The following is the account given by PLISSON* of the circumstances under which life is destroyed by drowning, and which tallies pretty closely with the observations made by Dr. GOODWYN† in drowning brutes. Allowing a little for the usual French colouring, he has given a very interesting description, part of which he collected from the statements made by drowning persons, after their recovery, of the feelings they experienced during the short time they continued conscious.

" Engulfed in the water, either by his own act or otherwise, the drowning man puts forth all his energies to gain the surface and breathe. Mastered by the dreadful thoughts which pursue him, the wretched creature, urged by despair to self-destruction, throws himself into the waters ; but scarcely has this momentary determination been effected, than the habitual instinct of self-preser-

---

* Essai historique et thérapeutique sur les Asphyxies. Paris, 1826. Second edition.

† Connexion of Life with Respiration. London.

vation, which Providence has wisely implanted in
animals, regains its natural supremacy, and the
wretch employs all his strength to elude the death
which he had just solicited as a relief from his
woes. If he cannot swim, he struggles, sinks,
rises several times to the surface of the water;
but soon his presence of mind and powers abandon
him ; he ceases to defend himself against an enemy
which he cannot overcome, and reappears no more.
Then the pulse is weak and frequent ; there is a
sort of dizziness in the head, singing in the ears,
oppression at the chest, small inspiratory and ex-
piratory movements, a certain quantity of air escapes from the lungs in the form of bubbles,
which burst on the surface of the water. The
distress increases, the pulse becomes more feeble ;
new efforts are made by the drowning person to
escape the agony of which he is the prey : a fresh
quantity of air escapes from his chest, and is replaced by water, which passes into the mouth,
and thence into the windpipe and its branches.
The skin becomes livid at different spots ; the eyelids, the lips, the back, and the nails assume a
bluish hue ; at the same time the face and neck are
bloated ; the tongue acquires a larger size, but
rarely protrudes beyond the teeth, which are very
often strongly clenched together ; the pulse stops
by degrees, and the belly distends. A foam,
which some time after death becomes bloody, escapes from the mouth and nostrils ; the body

cools ; the limbs stiffen, the sphincters relax ; and there ensue total loss of sensation and motion, stifling, and death. Being under water continuously for three minutes and a half, and sometimes even less, is sufficient to produce stifling."

The cause of death from drowning is not, as formerly supposed, from the lungs being filled with water instead of air ; whence, in days of yore, it was the practice either to put the person into a barrel, with its ends knocked out, and roll him briskly about, so that the water might be dislodged by the agitation, or to hold him up by the heels, so that the water might drain out, as from a vessel turned upside down. The person is destroyed by no fresh air being received into the lungs, as, directly he is completely under water, the mouth and throat being filled with it, the water irritates the top of the windpipe, and the epiglottis closes firmly and prevents its further passage. There remains then only the air already contained in the lungs, and, as this becomes spoiled by the passage of the blood through them, it expands, and for a moment pushes up the epiglottis, a small quantity of water rushes in, and the epiglottis closes as fast as before : this may be repeated once or twice, but the epiglottis always closes, and the person is as completely destroyed from lack of fresh air as if he had fallen into any place strongly impregnated with carbonic acid gas, or had been hanged or strangled.

It is a very difficult matter to determine the length of time which has passed between the time when he was completely under water and that at which he was fished up. Hence has arisen the great difference of statements as to the time at which a person can be recovered under these circumstances The cases related of persons who have been restored after having been under water days, and even weeks, must be considered fabulous.

" When the person has been completely under water," says PLISSON, " three minutes and a half, and sometimes less, are sufficient to produce stifling. As to actual death, which commonly ensues when the drowning person is not immediately succoured, it often does not take place for some time : and, indeed, if faith could be put in some accounts of persons beneath the water, the living principle has been preserved for many weeks whilst they remained there. But these accounts must be considered apocryphal, when it is mentioned, that of fifty-three drowned persons restored to life, in the space of one year, by the Amsterdam Society, but one single person was beneath the water a whole hour; very many only a few minutes; a small number a quarter, and some half an hour.

KITE mentions, from the records of the Humane Society, one only out of six hundred successful cases in which the person was in the water

U

an hour; "and," he adds, "he floated on the surface of the water during the whole of the time;"
it may therefore be very much doubted whether
this be of any value.

DR. DOUGLAS has within the last few years
mentioned * the case of a man who had been
under water fourteen minutes, and showed no
signs of recovery till means were employed for
the purpose. SMETHURST gives the history † of
a girl two years old, who had been ten minutes
in the water; but there are doubts of the whole
body having been beneath it. WOOLLEY mentions ‡ one case in which the person had been
completely under water for five minutes. TAYLOR
states §—"that when the mouth is so covered
that air cannot enter, asphyxia (stifling) super-
venes in the course of one or two minutes at the
farthest;" and by comparison of various cases,
he further adds—"it would appear, from these
and other observations, that asphyxia is probably
induced in most individuals in the course of a
few seconds, and that at the farthest it occurs
in from a minute to a minute and a half."

When a drowning person has been removed,
after having been beneath the water for a few
minutes, he is cold, insensible, without pulse,
motionless, but the limbs are lax; his eyes are

---

* London Medical Gazette, vol. xxxi.
† Lancet, 1839-40, vol. ii.   ‡ Lancet, 1841-42, vol. i.
§ Manual of Medical Jurisprudence, 1844.

closed, his mouth sometimes firmly closed, sometimes gaping and covered with a clammy froth,
the face more or less livid and puffy, and the
body swollen and soaking with water. In carrying him, great care should be taken that his head
do not drop below the trunk, or his danger will
be increased by the blood pouring into and loading the brain. If placed on a board, his head
should be a little raised. and carefully propped,
so that it do not loll on either side. If carried
by hand, he should be borne by six persons forming a sort of litter by joining hands, four of whom
should bear the trunk, with the shoulders a little
raised, whilst the other two carry the legs. Another person should be solely occupied in carrying
the head, keeping it a little higher than the
shoulders, and taking care it do not roll about.
This number of persons is quite sufficient,—more
are only in each other's way, and in jostling are
liable to throw the drowned person down.

Treatment.

Before describing the means to be adopted, so
soon as the person has been removed from the
water, I cannot avoid quoting as a prologue,
and as an encouragement to exertion under the
most unfavourable circumstances, the following
excellent observations of TAYLOR :—" We are
not to allow ourselves to be influenced, in the
treatment of the drowned, by the shortness of

the period at which death must commonly take place : for it is possible that two individuals may be drowned under the same circumstances, and treated, on removal from the water, in the same way ; and yet the means of resuscitation will be effectual in one case, while they will totally fail in the other.   It ought to be borne in mind that the susceptibility to the restoration of life may be different in the two subjects : were not this the case, it would be impossible to explain why, under the most judicious treatment, every effort will fail in restoring animation in a subject which has been submerged only a few minutes, while the same means will perfectly succeed in resuscitating another subject, which may have been submerged more than twice the period."

The great objects in the treatment of drowned persons, are to re-excite the circulation and breathing, which are suspended, and to restore the temperature of the body.   If any house be near, the person should be immediately carried to it, quickly stripped of his wet clothes, and put into bed between the blankets, which have been warmed with a pan of coals whilst the clothes are removing.   If at a distance from a house, and the sun be bright and powerful, he had better be at once stripped, and laid, fully exposed to it, on such dry clothes as the bystanders may be induced to give up, and spread out for him.   The body should be wiped dry, the mouth cleared of the clammy froth, the head and shoulders a little raised, heat applied to the pit of the stomach and soles of the feet, and rubbing with coarse flannel, or a jacket employed incessantly over the body and limbs, but especially over the chest. The heat may be furnished by bottles of hot water, by hot bricks, sand or ashes in woollen stockings, by a pan of warm coals, taking care it be not too hot, by warm grains if near a brewhouse, or even by immersion in a warm bath, if it can be had, taking care, however, that the head do not drop in. In short, anything and everything from which warmth can be obtained may be brought into use. Endeavouring to get air out of and into the lungs, should also be tried by pressing the chest and belly as already directed (p. 280), but it should not be persisted in if it interfere with the rubbing, which is the principal thing to be relied on.

As the rubbing may require to be continued for some hours, it will be necessary that there should be relays of assistants to keep it up effectually. In DR. DOUGLAS's case above mentioned, and which is one of the most remarkable on record, no sign of revival appeared till the rubbing had been continued *eight hours and a half from the time of the accident.* If this be not encouragement to perseverance, it will be difficult to find any like it.

It must be particularly remembered, that the room in which the drowned person is placed be spacious and airy as possible, and that the number of persons in it should be as few as may be, consisting of those only actually employed in the patient's restoration, and one or two others to render them assistance. Even the relay of rubbers should wait in a neighbouring room till their services be required, and then those who are relieved should leave the room. The greater number of persons in the room, the more impure the air is rendered by their breathing, and the probability of the patient's recovery is thereby diminished.

So soon as the warmth and rubbing begin to take effect, a slight convulsive movement of the chest is observed, and the air is drawn into the lungs with a sigh or sob. The assistant's efforts must be now continued most diligently—life is not extinct—but the least relaxation on their part may cause "the silver cord to be loosened." Another sigh or sob follows in a few minutes, the jaws begin to relax, a slight flutter may be felt at the heart, and the pulse may be found beginning to beat very feebly. By degrees breathing becomes more decided, slow at first, but afterwards increasing in quickness, and the heart's action becomes more distinct. As soon as the mouth can be got open a little, warm tea or weak wine and water should be carefully given, taking especial care that it be put far back into the throat, or it may be poured into the windpipe, and do serious mischief. By putting the spoon far back, it for the moment presses down the epiglottis, and so the chink of the windpipe is protected. Sometimes the person slowly and quietly recovers his senses, at other times he will screech out for some hours, as his consciousness seems to recover itself, as if it were resumed from the moment when he had lost it, and he still considered himself in the act of drowning.

When the breathing, and the circulation of the blood, and the warmth of the body have been restored, it must not be supposed that the person requires no further care. He must be watched carefully for some hours ; and if there be continued insensibility, or a recurrence of it and convulsions, it may be presumed that the brain is labouring under the effects of the foul blood which has been some time stayed in it ; and then it is right to take away some blood—by cupping from the back of the neck is best ; but the quantity taken must depend on the effect it produces, or fatal fainting may ensue, and the remedy become as bad as the ailment for which it is employed. The person should be kept in bed, and warm drinks given, a little at a time, so as to excite the skin to perspiration, and thus assist in ridding the blood of more carbon.

HANGING.

When a person is hanged, he is most commonly destroyed by the rope or handkerchief pressing so tightly upon the windpipe, below the voice-organ, that the air cannot pass into or out of the lungs, and he dies under precisely the same circumstances as when the epiglottis is closed by drowning, or by his being plunged into very foul air. If, however, the rope be very high up in the neck, the strength of the voice-organ may prevent the windpipe being so closed that air cannot enter the lungs, and accordingly he may continue breathing for a time, but is at last destroyed by apoplexy, the blood not being able to return from the brain.

A great deal has been said about dislocating or breaking the neck in hanging; and the drop employed in this country for executions was formerly considered a most admirable contrivance for this purpose, and it was held that thereby the criminal's sufferings were more quickly terminated. The celebrated French surgeon, LOUIS, was curious to ascertain how it happened that some persons who were hanged died immediately they were turned off, whilst others died more slowly; and on inquiring from the executioners of Lyons and Paris, the former of whom was the quicker operator of the two, he found that he broke his patient's neck, whilst his brother of Paris only compressed the windpipe so tightly that the air could not pass through it. This breaking or dislocating the neck may probably have been effected by the Lyons finisher of the law, if he practised like the executioners of Spain, who, the moment the unhappy wretch is thrown from the ladder, jumps upon his shoulders, like an incarnate fury, and thus his weight suddenly thrown on may cause a dislocation. However, in the bodies of executed persons which I have examined, I have never seen an instance of such dislocation or breakage of the neck.

When a person hangs himself or is hanged, it does not appear that he suffers much pain as the noose tightens round his neck, but merely a sense of stupefaction and numbness, which ceases only with loss of consciousness. Sometimes he has a ringing in the ears, an appearance of blue light, or a kind of flame which gradually sinks into darkness. Death very soon follows, and after the person has been hanged but a very few minutes it is rarely possible to recover him. The countenance and lips generally appear bloated and of a livid colour, the jaws are half closed, the mouth frothy, the tongue sometimes protruded and swollen, the eyes red, swollen, and starting from their sockets, the hands tightly clenched and, as well as the nails, livid.

Treatment.

Immediately you find a person hanging, cut him down, and take off the rope or handkerchief or whatever he has been suspended with, from his neck. Do not waste precious time by running for assistance, but yourself act at once, for very rarely any good is done in these cases even after

hanging only a few minutes. The body should
be stripped, dashed with cold water, blood should
be taken from the arm, and stimulating liniments
rubbed perseveringly on the chest.

A very remarkable instance is related by Dr.
PLOTT * of a woman who, in the reign of
HENRY VI., was hanged and remained so a
whole night, but was alive next morning and
was pardoned. PLOTT states her escape depended on her voice-organ being converted into
bone, and consequently that the rope which had
noosed about it did not prevent her breathing.
This is at least a very marvellous narration, notwithstanding PLOTT is a very trustworthy authority. Another very curious story is mentioned,
by Dr. MAHON † of a butcher who was hanged
for murder at the Old Bailey in the beginning
of the last century. He was rich enough to
bribe a young surgeon to make an opening in
his windpipe, below where the rope would come,
before he went to execution. He was supposed
to have attempted to commit suicide, and his
execution was hastened. After hanging the
appointed time he was cut down and quickly
conveyed to a neighbouring house, where he was
bled and other means of restoring life employed :
he opened his eyes, sighed deeply, but in a few
minutes died.

---

* Natural History of Staffordshire.
† Médecine Légale et Police Médicale, vol. iii.

CHOKING BY THINGS GETTING INTO
THE GULLET.

PEOPLE are sometimes choked and killed in a few minutes when eating quickly and carelessly. This happens in two ways.

First, by a large piece of meat sticking in the throat and preventing the air passing into the windpipe, of which I knew an instance in a man who was eating leg-of-beef soup for his supper, rather greedily it may be presumed, as he got into his throat a piece of meat a couple of inches long and about three inches round. He left the table immediately, went out at the street-door, and about a quarter of an hour after was found dead. It was supposed he had died of apoplexy, but on examination of his body this large piece of meat was found in his throat, and there could be no doubt it was the cause of his death. Had the accident been suspected, the meat might have been easily pulled out with the finger and thumb, as it was quite within reach. This was of course a very extreme case. But hasty eaters are often liable to great pain and distress in bolting large pieces of food, which for a time stick in the gullet, sometimes higher, sometimes lower. This

may usually be overcome by taking large
draughts of water and making great efforts to
swallow; the quantity of water distends the
gullet above the lodged food, alters its position,
and both water and food pass into the stomach
with a sudden jerk. If this do not succeed, a
medical man must be found to pass an instrument called a bougie into the gullet and push the
lodgment down.

The second way in which a person may be choked
requires no large piece of meat, and may happen
to any very cautious eater, if the meat he chews
be stringy. In this case the meat may be
chewed sufficiently small, but two pieces of it
remaining attached, like a chain-shot, one piece
is swallowed whilst the other is entangled in the
teeth, and the consequence is that the string,
connecting the two, shuts down the little trap
at the top of the windpipe, and stops the breathing; and the greater the effort to swallow the
more tightly the trap is shut down. It is,
therefore, always advisable when any choking
occurs during eating to thrust the finger and
thumb as far back into the throat as possible,
and if there be anything there, to pull it out
forthwith.

Pieces of bone, more commonly fish-bones, are
sometimes swallowed at meal-time, or a woman
will occasionally swallow two or three pins, when
guilty of the foolish practice of holding pins in

her mouth, instead of fixing them in any convenient part of her dress.   Sometimes both bones
and pins pass down the gullet into the stomach,
either gently scratching the gullet as they go along
or not even producing any immediate inconvenience.   But at other times the bone or the pin
lodges, and the points running into the gullet,
there it remains.

These are rarely dangerous accidents, but they
are very distressing, cause much pricking in every
attempt to swallow, and are sometimes accompanied with violent cough and vomiting.   Occasionally, without anything being done, after a few
hours or a few days the bone or pin, by some accidental movement, changes its position and passes
into the stomach.

Generally after a fish-bone or pin has been
swallowed and passed into the stomach, sometimes
when it has been vomited up or been removed by
assistance, there remains the feeling of the throat
being scratched, and of pricking when anything
is swallowed, as if the intruder were still there.
Usually it is only the scratch which remains, and
we are not therefore wrong in consoling the patient
by telling him that it will go off in two or three
days.   But we may be mistaken about this, for I
have known an instance in which a fish-bone was
swallowed and pulled out in the course of a few
hours ; but the distress and difficulty in swallowing
continued, in consequence of which a bougie was

passed on the fifth day, and readily descended into the stomach, so that it seemed quite sure there was no obstruction. But on that very same evening a violent fit of coughing came on, and a second bone was thrown out, immediately upon which the relief was complete.

But at other times it is shot up by vomiting, or by a violent fit of coughing. Sometimes if the bone or pin be near the top of the throat, it may be got out by pushing the finger far down, and hooking it up with the nail. But if below the reach of the finger, the best thing to try for immediate relief is to take some crust of bread or some hard apple into the mouth, chew it coarsely, get down two or three mouthfuls without swallowing it completely, and then to swallow quickly three or four full gulps of water, which acts like a rammer to the bread, and forcing it against the bone or pin, not unfrequently carries it down into the stomach, and there the matter ends.

If this do not answer, the doctor must be applied to, but if there be none to be had, these attempts should be repeated.

Children sometimes swallow pieces of money, buttons, shells, nails, or anything else they can put into the mouth and get down. The parents are generally excessively frightened and anxious to get them away, on which account it is common to purge the child again and again till the intruder is expelled. These accidents are very rarely of any consequence ; the money or other substance usually accompanies the food through the bowels, and is in due time, sometimes sooner, sometimes later, discharged with the motion, and the more solid the latter is the more likely it is to entangle the money; and therefore purges are, on the whole, best avoided. The celebrated deceased sculptor CHANTREY, indeed, proposed feeding on suet-pudding for a few days as a good trap for the money or other substance swallowed, and the notion is not by any means a bad one.

Dr. MONRO, however, mentions the case of a boy who got a halfpenny fixed fast in his gullet, where it remained for three years, when the boy died of consumption.

CHOKING BY THINGS GETTING INTO THE VOICE-ORGAN, OR INTO THE WINDPIPE.

It has been already shown how carefully the chink of the voice-organ is protected by the epiglottis from either fluid or solid substance getting into it. But occasionally it happens that a person, whilst drinking or eating, thoughtlessly attempts to speak, for which purpose he empties. his lungs or he may merely empty his lungs without attempting to speak, and thus forcibly pushes up the epiglottis from the chink of the voice-organ; immediately this happens, a portion of the fluid or solid, whichever may happen to be at the back of the mouth, slips in and causes violent spasmodic cough with a feeling of being strangled, and the cough continues, the person gasping also for breath till the intruder be expelled; often, indeed, the cough does not cease for some minutes, and becomes exceedingly distressing. This is the condition in which a person is placed when "something has gone the wrong way," as it is commonly called.

If the substance, which may be a small stone, a piece of shell, a fish-bone, a pin, or even a tooth-brush bristle, get *fixed in the voice-organ itself*, the symptoms are most violent, and unless it can be got out the person is usually destroyed in a few days by the irritation, inflammation, coughing, and difficult breathing which are produced.

Treatment.

This accident scarcely admits of any other relief than a surgical operation, which none of my unprofessional readers would be justified in performing. The only hope under these circumstances is, if no surgical assistance can be had, that the violence of the cough may dislodge the substance and shoot it out.

The only thing justifiable for a person not a doctor to do is, that one who has a long forefinger should pass it as far back as he can into the throat, and endeavour to get it behind and beneath the epiglottis, with the slender chance of finding the substance either lying across the top of the chink of the voice-organ, or partly projecting from it. In either case he may perhaps be able to dislodge it into the gullet, or hook it out with his finger nail ; and he is more likely to succeed in this way when the substance lies upon, and not partly in, the chink of the voice-organ. The chance of success in this way is very slight, and still more so according to the smallness of the substance which has slipped in.

The substances, however, which slip through the chink of the voice-organ more commonly pass down the windpipe, descending even below its division, and getting into one or other of the two great branches from it, which run to the two lungs. The form, size, and weight of the thing which has thus got into the windpipe generally, though not always, determines whether it shall move up and down in expiration and inspiration, or whether it shall continue fixed in one spot. Small stones, or beads, peas, cherry-stones, and the like, are things which most commonly slip into the windpipe of children, and these can, usually at first, be felt shot up to the voice-organ at every expiration, and dropping again at each inspiration. But if the substance be a piece of money, or a nail, it in general lodges either somewhere in the great trunk of the windpipe itself, or in one of its great branches, and changes its place rarely and with difficulty. The coughing in these cases is less constant and distressing ; but it comes on in violent fits at intervals, and in one of these the substance is sometimes shot out with great force. In this condition there is much less danger than when the substance is lodged in the voice-organ, because the windpipe is less irritable, and will even by degrees accustom itself to its ugly tenant, more especially if it have got lodged in one of the branches of the windpipe ; and sometimes after a few days the only inconvenience is a dry cough.

The length of time bodies will remain lodged

in the windpipe is very variable. A very remarkable instance is on record, in which a girl of
nine years old got a piece of chicken-bone into
her windpipe, which was not thrown up till she
was twenty-four years of age. In another case,
under the care of one of my medical friends, a
plasterer threw up a lath-nail, which had slipped
into his windpipe more than twelve months before.
Both these persons died some few years after with
diseased lungs, which is the usual consequence of
a foreign body retained any length of time in the
windpipe or its branches, on account of the slow
kind of inflammation which they excite in the
lungs.

Treatment.

If the substance be a light one, and can be
felt moving up and down the windpipe continually, as the air is thrown out or taken into
the lungs in breathing, or if the substance be
heavy and can hardly be perceived to move when
the person breathes, there is good hope that,
without any actual surgical operation, the escape
of the body may be rendered easy, and by a little
patience and perseverance, that it may be completely got rid of. Two very striking instances
of this kind are related, and the first of them, as
being a very simple remedy, is the one I should
advise to be pursued. A Highland shepherd,
whilst mumbling a small bullet between his teeth,

x 2

unfortunately had it escape from them and slip into the windpipe. He coughed incessantly for two hours, after which he had slight inconvenience beyond a little occasional dry cough, till the middle of the following day, when he was attacked with shivering, head-ache, and deep pain in the right side of the chest. The shivering and head-ache ceased, but the pain continued, and he was excessively drowsy. On the evening of the third day he was seen by Dr. MACRAE, who, being satisfied of the lodgment of the bullet, "directed the man to be strapped securely to a common chair, that he might be easily suspended from the rafters of the roof with his head downwards, in order that his chest might be conveniently shaken by a rapid succession of sudden smart jerks, and that the weight of the bullet might favour its escape from its seat in the lungs. He was kept depending as long as he could endure such an uncomfortable position, and then placed in the horizontal posture for a few minutes to rest. When sufficiently recruited, he was hung up again. Upon being taken down the first time, he described the pain in his breast as moved nearer to the top of his chest, and during the third suspension he joyfully exclaimed, ' Thanig à, Thanig à!' ('It has come, it has come!' in the Gaelic language) immediately after a smart shaking, and a few convulsive retching coughs, and spat the little bullet from his mouth.

The diameter of it is three-eighth parts of an inch."*

The second case is that of a celebrated engineer, which excited considerable interest in London. Whilst playing with his children a half-sovereign slipped into his windpipe, and was followed by the usual symptoms. On the sixteenth day after the accident he made an attempt to get rid of the coin " by placing himself in the prone position, with his breast resting on a chair and his head and neck inclined downwards, and, having done so, he immediately had a distinct perception of a loose body slipping along the windpipe. A violent cough ensued. On resuming the erect posture he again had the sensation of a loose body moving in the windpipe, but in an opposite direction, that is, towards the chest." The experiment was repeated six days after, more completely; " he was placed, in the prone position, on a platform, made to be moveable on a hinge in the centre, so that one end of it being elevated, the other was equally depressed. The shoulders and body having been fixed by means of a broad strap, the head was lowered until the platform was brought to an angle of about 80 degrees with the horizon. At first no cough ensued, but on the back having been struck with the hand, the patient began to cough violently ; the half-sovereign, however, did

* LISTON's Practical Surgery.

not make its appearance. This process was twice repeated with no better result; and on the last occasion the cough was so distressing, and the appearance of choking so alarming," that it was not thought right to proceed further. Two days after, the windpipe was opened by a surgical operation, but the money could neither be felt nor got out. He was therefore left alone for ten days to recover the effects of the operation, and was then placed again upon the moveable platform in the same position already mentioned, the back was struck with the hand, " cough followed, and he presently felt the coin quit the bronchus, striking almost immediately afterwards against the front teeth of the upper jaw, and then dropping out of the mouth."*

---

* BRODIE, in Med. Chir. Trans. vol. xxvi. 1843

THINGS PUSHED INTO THE NOSE
AND EARS.

CHILDREN often amuse themselves with poking
things with which they are at play, into their
noses or ears. If peas, beans, or any other
seed or substance be thrust in, which swell as
they moisten, no time should be lost in getting
them out, otherwise, as they enlarge, they become more firmly fixed and more difficult to be
removed, are attended with great pain and suffering, and may even cause dangerous consequences.
Hard substances, as shells, which remain unchanged in bulk by moisture, are of less consequence, and may remain some days without
causing much inconvenience ; and often drop out
of themselves.

If the pea or shell be *in the nostril*, the child
should be made to draw his breath in deeply,
and then stopping the other nostril with the finger,
and closing the mouth firmly, to snort forcibly
through that side of the nose in which the substance is lodged. If this be done soon after the
accident, two or three efforts usually shoot the unwelcome lodger out. But if this do not succeed,
the nose must be tightly nipped with the finger
and thumb above the pea or shell, so as to pre-

vent it getting further in, and then the eyed end
of a bodkin or probe, having been a little bent,
must be gently insinuated between the bottom of
the nose and the substance, and when introduced
sufficiently far, must be gently used as a hook to
bring it down. Pushing it back into the throat
should not be tried, as not unfrequently so doing
only fixes it the more firmly. If a doctor be
within reach, it is better at once, if the substance
cannot be snorted out, to take the child to him,
as he will be able to manage the matter better
and more readily the earlier he is applied to.

It is of much greater consequence when any
thing has been pushed *into the ear*, as, though
the passage is short, its nearly circular form and
smooth surface more readily permits it being
quickly thrust almost to, or even quite down to
the drum of the ear. The passage is also so
narrow, that it is difficult to get in either the
end of a bodkin or eyed probe between the substance and the ear-passage, and not unfrequently,
indeed, it is pushed farther in. If it were advisable to attempt the early removal of any
swellable body from the nose, it is ten . times
more so when one such is lodged in the ear-
tube, nearly the whole of which being very unyielding, the agony which the ·swelling body
produces by its enlargement is extreme   The
doctor therefore should be immediately sought
for. No syringing with water or any other

fluid should be resorted to, as it will excite the pea to swell, and increase the mischief; and dry heat alone must be employed.

If, however, a hard body, as a shell or button or bead, be pushed into the ear, syringing with water may be used with advantage, as, if the water pass in any way between the hard body and the ear-drum, it will not unfrequently force it out. The head should be laid down, so that the ear, in which the hard body is, be undermost, and in this position the water should be thrown up with the syringe, the nozzle of which, however, must be held at some little distance, and not put into the pipe of the ear, or it will prevent the hard body dropping out. Whilst the head is thus laid on side, I have known the offender ousted by a smart box on the other ear.

Poking the ear with a bodkin or probe, if there be doubt of any thing being there lodged, and, indeed, if it cannot be distinctly seen, should always be avoided. Instances have occurred in which death has ensued, from persevering attempts to get out substances, which upon examination were found not to have been there at all.

Insects, sometimes, though rarely, get into the ear, and cause much inconvenience even if they do not sting and produce further mischief. The best mode of proceeding in such case is to fill the ear with sweet oil, which will kill the animal by

stopping up its breathing pores and generally
floats it out. But if it be not thus dislodged, it
must be washed out with a syringe and warm
water.

Sometimes, from want of cleanliness the tube
of the *ear becomes loaded with wax*, which dries
on the surface next the ear, and contracting and
cracking as it dries, and allowing the air to get
between it and the air-passage, causes all sorts
of odd noises, from the singing of a kettle to the
roaring of a torrent, with occasional sharp sounds,
like the report of a pistol. Sometimes these
annoyances are accompanied with pain, but more
commonly not. The worry and confusion, however, are distressing beyond conception, except
by a sufferer, and completely incapacitate for any
employment. Occasionally the hard wax irritates, and sets up inflammation with its attendant
ear-ache, which ends in matter being formed,
and the waxen plague being pushed out.

In either of these cases, syringing with simple
warm water should be employed. Simple warm
water is better than soap and water, which generally irritates, as not rarely the scarf-skin
comes away with the wax and leaves bare the
true skin, raw and tender, and often smarting with
the warm water only ; but it smarts severely, and
is much irritated and excited to discharge by the
application of soap and water. As the wax is
always more or less hard, it is the best plan to

fill the ear with bread and milk poultice over-
night; and then washing out with syringe and
warm water next morning, generally brings away
also the collected wax without difficulty ; although,
whilst unsoftened, it often will not move at all.
When the ear-tube has been cleared and dried
with some soft linen, a small quantity of warm
oil should be dropped into it, and kept in with a
little wad of cotton laid against, but not pushed
into the ear-tube.

Poking the ears with bodkins and ear-pickers
is an abomination, and often repays the practiser
of it with a smart ear-ache.

THINGS IN THE EYE.

THINGS of various kinds occasionally *get into the
eye* as it is commonly called, not meaning thereby
that they get into the eyeball itself, though this
indeed sometimes happens, but only that they
get within the eyelids, between them and the ball.
A small fly may dash in, or road sand, small
piece of straw or soot, or any other small body
may be driven in by a gust of wind ; or whilst
grinding a steel instrument a delicate fragment
may be thrown up by the wheel into the eye.

If the body be soft, or if it have not been darted in with much violence, it rarely fixes on the globe but is quickly transferred from it to the inside of one of the eyelids, generally of the upper, so that it is not at first perceived, and its presence only supposed in consequence of the uneasy feeling and the free discharge of tears, which are secreted in great abundance as a natural effort to wash the offender out. If however it be driven with violence, it beds itself more or less deeply and firmly in the surface of the eyeball and can then be seen as well as felt when the lids are opened.

When a person feels that something has got into his eye, the first thing he does is to close the eyelids violently, open them partially, close them again and compress the eyeball in every way the muscles will admit, or in common but very expressive language "screw the eye up" for the purpose of squeezing out the unwelcome intruder; whilst at the same time he rubs the eyelid smartly with his knuckle. This is often followed by a gush of tears which floats the little body from its lodgment and entangles it in the eyelashes whence it is easily removed. Not uncommonly however this proceeding is unsuccessful, and if the intruder instead of lying loosely on the surface of the globe or eyelid, have fixed itself, the more rubbing the less likely is it to be detached; therefore under any circumstances it is better to proceed in a different manner.

The most simple plan is, to keep the eyelids closed and then gently pass the finger over them from the outer to the inner corner, which is the natural course of the tears in their passage from the gland secreting them on the upper outer part of the orbit, to the little passage by which they run into the nostril. After thus passing the finger a few times the little substance is often found at the inner corner of the eye and may be wiped out with a handkerchief or with the finger. Another and very good method, if the lodgment be either on the eyeball or within the upper lid, is to take hold of the eyelashes of the upper lid and lift it completely over the lower, the lashes of which being thus interposed between the upper lid and the eyeball serve as a delicate brush and entangling the substance bring it out with them, when the upper lid recovers its place. A third proceeding is to lift up the lid, and insert a piece of blotting paper between it and the globe, and giving it a gentle sweep the intruder is caught by and removed with it. Should neither of these methods however produce any relief; if the uneasiness and the watering of the eye continue, then it is better to ascertain the precise situation of the intruder so as to act more efficiently on it. For this purpose, with the finger gently draw down the lower eyelid which will completely expose its inner surface and the whole lower part of the eyeball, both of which are to be examined.

To look at the upper lid, which requires being
turned inside out, is more difficult, but may be
done with gentleness and a little dexterity in the
following way : with the finger and thumb of one
hand take hold of the lashes of the upper eyelid and
pull it forward from the globe, then with the other
lay the blunt end of a bodkin or of a small steel
netting needle upon its outer surface and press down
gently at the same time lifting the lashes towards
the eyebrow. This turns the inside of the lid
out and it may then be well examined.

If the little substance be found unfixed, it may
be either gently sucked off with the mouth or with
a piece of blotting paper ; but if partially em-
bedded it will need brushing rather firmly with
the end of a bodkin wrapped round with a small
piece of lint, and thus removed.

Occasionally it happens that the intruding substance is got rid of unconsciously in rubbing the
eye, but having wounded and irritated the surface
of the eyeball or lids, the irritation may continue
for some time, and in some cases produce super-
ficial inflammation of the ball or lids. Generally
however the uneasiness subsides, but if it continue and increase, it is most likely that the cause
of irritation is still remaining

ON THE DRESS, EXERCISE, AND DIET
OF CHILDREN.

———◆———

ALTHOUGH this subject do not strictly fall within the limits of this little work, yet I am induced to make a few observations upon it, for the purpose of setting before parents, especially mothers, the grievous folly, to call it by the mildest term, of the general mode of dressing, exercising, and dieting children, more particularly girls, whose health is ruined and their lives shortened by the unnatural, I had almost said atrocious system, to which in youth, if not even in childhood, they are subjected for the improvement of their figure and gait, as it is called, though in reality it is only for the purpose of making girls into women before the proper time.

Dress.

Boys in general are soon properly clothed: it is the custom that little master be early slipped into a pair of trousers, and have his chest completely protected by the little tunic which reaches above his collar-bones. He may wear . a belt, indeed, for ornament, but it forms no tight girth about his chest. In fact, his chest

and limbs have as free play as if he wore no clothes
at all, and the result is, that he grows up tall,
straight, and with a well-formed chest, which
allows the free play of his heart and lungs. Rarely
during his youth is this happy condition for his
health interfered with : he may arrive, indeed,
at the dignity of stand-up collar and coat, sooner
or later, but scl ool discipline does not permit
screwing up his waist with a tight belt, to give
greater apparent breadth to his shoulders, a proceeding which some silly fellows adopt after they
have reached years of discretion !—fortunately,
however, their chest has by this time become so
completely formed, that their folly even cannot
do it much damage.

Now, let this sensible method of clothing boys
be compared with that in general adopted for girls.
The mischief often begins from the very cradle.
The upper part of the chest and shoulders are
rarely covered, except when the child is specially
dressed for going out of doors, as the shoulder-
straps of the several parts of the dress, instead of
completely covering and fairly resting on the tops
of the shoulders, as they should, are allowed to
slip below them, so that they cannot support the
dress, which is only kept at all in place and prevented dropping completely off by the waistband
or strings which surround the chest. The consequence of this is, that the shoulders and upper
part of the chest are continually exposed to

draughts of air, or to sudden changes of temperature, which cannot be avoided even in-doors, in
passing from one room to another; and thus is
laid the foundation of irritable lungs.

When the child begins to run about, the top of
her dress, though in some degree altered, being
still too wide to allow the proper resting of the
straps upon the shoulders, so as to keep it up,
the dress slips down first on one side and then on
the other, and to relieve herself from this inconvenience, and even to prevent the dress dropping
entirely, the child is constantly hitching up first
one shoulder and then the other, or even both at
once. Usually, however, both sleeves do not drop
down equally, and thus one shoulder becomes the
habitual hitcher, and the trunk in this action
being continually thrown to the opposite side, the
spine naturally bends that way, and hence very
frequently originates a crooked or curved spine,
which is usually first discovered by the medical
attendant whose advice is asked, on the mother's
attention being drawn to one shoulder being higher
than the other, and wishing to have it corrected.

The mode of preventing these fearful consequences is simple enough. The soundness of the
child's lungs, and the straightness of her spine, are
preserved by one and the same mode of management. *Dress the girl properly* by having the
frock at least as high as the collar-bones, and the
upper opening of the other parts of her dress only

Y

so wide, that the shoulder-straps can have a good
bearing, and not slip off the shoulders; the dress
will then be supported as it should ˙ be, on the
shoulders, and cannot slip down, consequently
there will be no hitching, nor any curved spine
from that most common cause of it, and the chest
will at the same time be protected from cold.

The next ill treatment to which a girl is subjected, is that of enclosing the greater part of
her body in bone stays, which, covering about the
lower two-thirds of her chest, and reaching as
low as the tops of the hip-bones, wide at the top
and bottom, but narrow enough in the middle,
with a broad steel or wooden busk in front, curved
forwards, but with its ends bent backwards, and
with a pair of stiff whalebones behind at the
lacing edges, besides a large array of other
smaller bones of the same kind, are so ingeniously
contrived as not merely to prevent the expansion
of the chest necessary for the proper performance
of inspiration, but actually to diminish its capacity at its lower or widest part; so that the poor
child has her chest really put into a vice, and can
only breathe very imperfectly, for the silly purpose
of preventing her waist being thick, which is the
real object of wearing stays.   Mothers, however,
sometimes delude themselves with the notion that
they thus encase their girls in steel and whale-
bone for the purpose of giving them support,
whilst in reality they are destroying their children's

health and constitutional powers by jamming up the great organs of respiration and circulation into a very much smaller space than is necessary for the due performance of their functions, and thereby disposing the lungs to that fatal disorder consumption, which, though commonly attributed to the variability of our climate, may, I believe, not less frequently, be traced to this most abominable custom of stay-wearing, and its attendant, tight lacing : for being early impressed with the notion that the elegance of their figure depends on the tightness of their stays, most girls greedily imbibe this pernicious habit, and screw their waists so tight, that they suffer constant distress for the sake of fashion.   The accom-

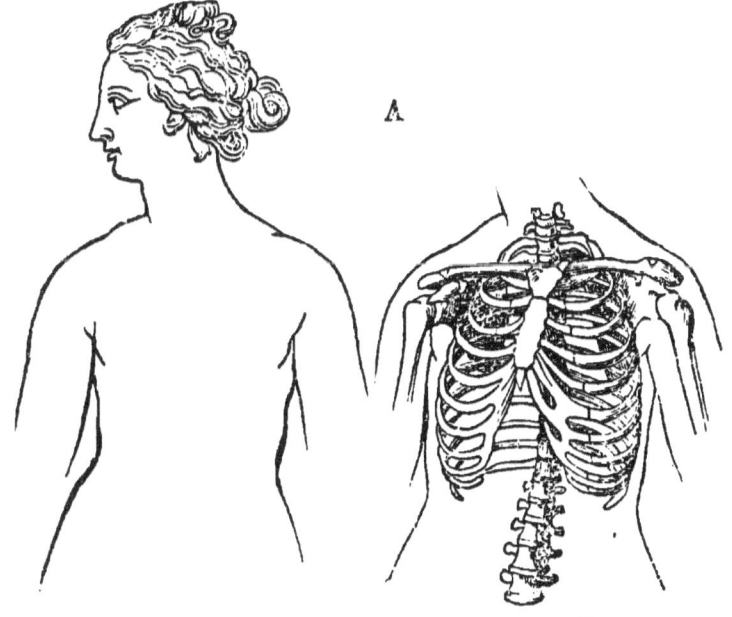

A.

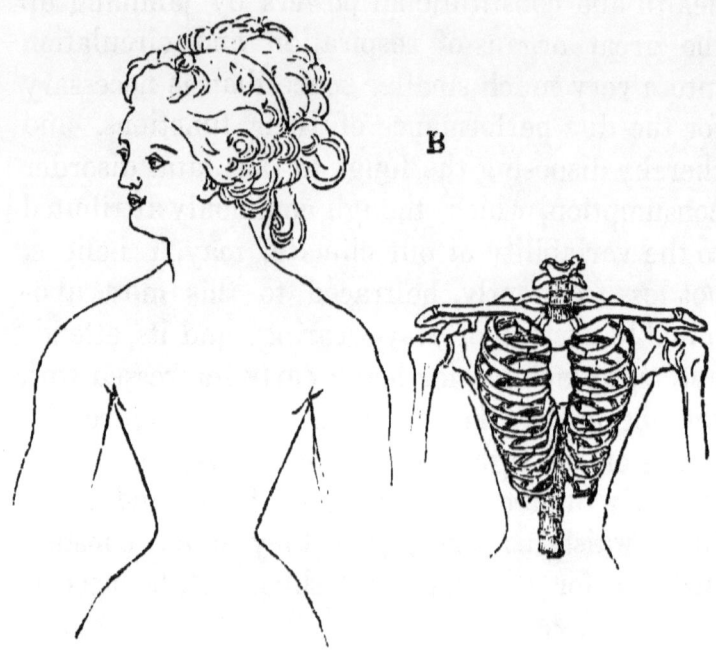

panying illustrations from a work of the celebrated anatomist SOEMMERING, will show (A) a
naturally-formed female chest, and (B) one which
has been ruined by tight-lacing.   This stay-wearing is as needless for the attainment of its object
as it is serious in its results ; for I have known a
few as well-grown and finely formed women as
could be met with, who never wore a bone stay
till eighteen or nineteen years of age, one of whom,
however, was silly enough so to encase herself
on her wedding-day for the first time.

Besides the more serious consequences to
which I have alluded, women are liable to various

other inconveniences from the abominable practice of tight-lacing ; for the stomach and bowels
being squeezed into an unnaturally small space,
not only by the contents of the chest being partially forced down upon them, but also by the
cavity itself in which they are contained being
lessened by the straitness of the lower part of
the stays, they are incapable of properly performing their duty, and hence the digestion of
the food being imperfectly effected, the consequence is that a long string often of very unaccountable and perplexing symptoms, commonly
called hysterical, are set up, and render the girl
miserable and incapable of the slightest exertion,
till at last she gets into positive ill health.
ABERNETHY is said to have likened the attempt
to reduce by tight lacing the natural capacity of
the great cavities of the body containing the
important organs of digestion, circulation, and
respiration, to that of attempting to put the contents of a quart pot into a pint ; yet odd as his
simile is, it is nevertheless quite true.   His advice to the parent who brought her daughter to
him for relief from a number of odd symptoms,
" cut her stay-lace," will be found in many cases
of complaining females to be a very excellent
as it is a very simple remedy.

One of my reviewers, in directing the attention
of his readers to what has been just said in reference to stays, says that there is " one little

fact omitted, which, if universally known, might probably have more weight" than what I have stated. I must confess I am not aware of the fact, though I am not disposed to argue its incorrectness. Be it however, as it may, his statement so closely concerns that which most persons are extremely careful of, namely, beauty of face, that it affords an additional and impressive reason against tight-lacing, of which I gladly avail myself. "If a foolish girl," says the reviewer, "by dint of busks and bones, and squeezing and bracing, secures the conventional beauty of a wasp-waist, she is tolerably certain to gain an addition she by no means bargained for, namely, *a red nose*, which in numberless instances is produced by no other cause than the unnatural girth obstructing circulation and causing stagnation of the blood in that prominent and important feature. Often in assemblages of the fair have we seen noses faultless in form but tinged with the abhorred hue, to which washes and cosmetics had been applied in wild despair, but alas, in vain! If the lovely owners could have known the cause, how speedily the effect would have vanished, for surely the most perverse admirer of a distorted spine and compressed lungs would deem the acquirement of a dram-drinker's nose too heavy a condition to be complied with."

Another point of great importance in the dress of children, too commonly alike neglected in

children of both sexes, is that of taking care, that
at all times and seasons of the year, the surface
of the body, particularly of the chest, should not
be suddenly cooled below its ordinary heat after
exertion in their play, during which most commonly children perspire freely, and in consequence their body-linen becomes completely
soaked with moisture, and as it quickly cools,
the child is wrapped, till it has dried, in a wet
cloth as completely as the foolish cold-water
sufferer is enveloped in his soaked sheet. This
is a very serious matter, as the surface being
chilled the perspiration is checked, and the lungs
have more blood sent to them to be freed ot
its carbon than is their due, and being thus
overloaded they are very often excited to inflammation.

Now, this is best guarded against by wearing a flannel waistcoat next to the skin; for
however sodden with perspiration it may be, it
cools very slowly, dries less quickly, and therefore after violent exercise the body is never
chilled as it is when linen is next to the skin.
Many persons have a foolish notion that the
warmth of the surface kept up by wearing
flannel next the skin is weakening, and fancy
they render their children more hardy by never
putting them in a flannel waistcoat. This, however, is nonsense : for as the preservation of an
equal temperature on the surface of the body

is at all times necessary for regularly sustaining the perspiration, which is almost, if not quite, as impoitant as breathing, in freeing the blood from its impurities, it is quite clear that the preservation of this condition should be most carefully attended to. Though I strongly advise flannel to be worn next the skin of the body at all times, and that delicate persons should in like manner cover their lower as well as their upper limbs, yet I do not mean that they should at all times wear flannel of the same thickness. The winter flannel waistcoat should be much thicker and warmer than that worn in summer, which may be as thin as can be procured, and will then not be found to add to the heat of the clothes.

Another bad practice to which most females of all ages are habituated is that of wearing at all seasons of the year, alike in the frosts of winter as in the heats of summer, thin cotton, almost gauze-like, stockings, in which there is not the slightest warmth, or thin silk stockings, which are but little warmer. This folly is for the purpose of setting off the foot and ankle to advantage, but the unwise mother in thus treating her girls forgets that she does so at the expense of cold feet, which are the result of the thereby diminished activity in the circulation of the blood through the feet, which in consequence of their distance from the heart, is naturally languid and ill needs any reduction of power. Care enough is taken that, on the approach of winter, boys be provided with some kind of worsted stockings or other, and most men make the same periodical change, and it seems utterly incomprehensible why females, who are generally less active than males, and therefore usually having a less active circulation, should be less protected from cold, to which their sedentary habits render them especially susceptible.

It is now the almost common practice for little girls to wear drawers, and a very good practice it is; and if it were always adopted by women it would be very praiseworthy, provided they do not show them.

If, then, you wish your children, girls especially, to have the best chance of health and a good constitution, let them wear flannel next their skin, and woollen stockings in winter—have your girls' chests covered to the collar-bones, and their shoulders *in*, not *out* of their dress, if you would have them straight; and do not confine their chests and compress their digestive organs by bone stays, or interfere with the free movement of their chest by tight belts or any other like contrivance if you desire their lungs should do their duty, upon which so mainly depends the preservation of health.

Exercise, as regards the bringing up of children, is of the utmost importance. It is not only necessary that they should have the free use of their legs, but also of their arms, so that all the muscles of their whole system should be brought into proper action, and, as a natural consequence, attain their proper bulk.   And when the child is tired with his efforts, he should rest himself in any position he finds most agreeable : for we may be sure he will put himself into such as will best relieve and rest the muscles he has tired by his exertions.

Now, boys are generally allowed to exercise themselves at their own pleasure, in running, leaping, or climbing, in cricketing, or hockeying, or kiting, in lifting, dragging, and throwing weights which they can master, and in various other exercises too numerous to be mentioned here ; and, if very active, will often change from one play to another, which does not call for the use of the muscles they have tired by their former game.   Only at last, when, to use a boy's expression, they are " dead beat," do they give up their laborious play and throw themselves down on a bench, or on the ground, in one of the many indescribable postures, with which, however, every man is himself perfectly well acquainted, and after a few minutes, change it for another, rolling about in all sorts of ways, till at last they have recovered themselves sufficiently to take to their legs again, and not improbably resume their violent plays.   In this way the child has the best chance of growing up to healthy manhood.

Now, let us compare this with the exercise generally permitted to girls, often even from a very early age. " Look here," as Hamlet says, " upon this picture, and on this ! "

As soon as a little girl has acquired the free use of her limbs, the first thing she is taught is " not to be rude and romp about, because such behaviour is not like that of a little lady ;" and thus at the very onset the child's natural and necessary activity is restrained, and she is put into an artificial training, by which she shall only move so much, and in such a way as is worse than useless ; and when she is tired by this wearisome proceeding, she may not throw herself down on the sofa, but must be seated in her little straight high-backed chair, with its bottom of scarcely sufficient width for her to sit upon, upright as if she were skewered, not with the slightest chance of her getting any rest, because she is obliged to call all the muscles of her spine into action to prevent falling off her chair.

In addition to this, she is subjected to the tender mercies of the dancing-master. I do not know whether that miserable invention, the stocks, is still in existence, but I well recollect the time when the poor girl had to stand for half an hour or an hour, with her heels together and her toes turned out, so that she should know how to stand properly—a most unnecessary torture, as nature has provided that women should, without any teaching, use this very posture, for without it they would stand very unsteadily, on account of the breadth of their hips. Then there was another abominable contrivance called a back-board, for keeping the shoulders well back, as it was called, by which the girl's arms were trussed behind her, in much the same way as the wings of a roast fowl. Next, the head and neck were to be borne in a particular way ; the chest was to be projected and the stomach pressed in, as it was delicately called. And all this posing was performed by the hands of the dancing-master till the girl, whatever her age, was set in the position which suited the fancy of the professor. Such was the way in which girls were pulled about by the dancing-master in former times ; whether it be still practised by him, or whether it fall under the province of the drill-serjeant, by whom in many of our fashionable schools young girls are taught to carry themselves and to walk, I know not, but it is as indelicate, as it is tormenting and useless to the poor girl ; and after all, by com-pelling her to move in an unnatural and forced position, the muscles of her spine are wearied, and the foundation of mischief laid.

If, as regards exercise, a girl is to have fair play, and if her parents desire she should be a well-formed and healthy woman, she must be al-lowed till at least ten or twelve years of age the same freedom from restraint, and to exercise her limbs as freely as boys do. I am no advocate for climbing ladders, or pulling ropes, or chair exercises, or any other of the unwomanly, violent, and ungraceful performances which have been dignified with the name of Callisthenic exercises, and which are too commonly advised for and practised by young women to repair the damage their form and constitution have suffered from restraint of exercise and ill training during their childhood.

Girls should be allowed to run about freely; they should use a skipping-rope, trundle a light wooden hoop, taking care however the stick with which they drive it change hands, so that the muscles of each side of the body be equally exercised, otherwise the spine will curve towards the most used, or the alone used side. They should be encouraged to play at battledore and shuttlecock, or trap-bat, and throw light balls. By one or other of these means, plenty of proper exercise may be undergone. And for the more especial object of strengthening the arms and the muscles of the shoulders, which will then do their duty, increasing the capacity of the chest, and setting the shoulders well back in their place, a pair of very light dumb-bells may be used, but never so long as to weary the arms. If dumb bells are not to be had, the same exercise may be as effectually performed by the girl taking a pound or two-pound weight in each

hand, grasping them and throwing them backwards and forwards, in the same way as in using
the bells.

When the child is tired she should be allowed
to sit down, not in one of those horrible narrow-
bottomed, straight, high-backed chairs, but on a
low chair in which she can sit comfortably, and
against which she can lean, and change her posture at pleasure.   She does not shift about from
one position to another from mere wantonness,
but that the tired muscles may be rested most
perfectly, and therefore she should have a seat
large enough to be able to do this, and not be
worried with continual exhortations to sit upright
and sit still.   If her system did not require this
change of posture for its better resting, she would
sit quiet enough.   I must confess I am a great
advocate for children resting themselves as they
may be disposed, on a sofa, or even on the floor.
Fatigue is never so quickly recovered from as in
the horizontal posture, or in any near approach
to it.   In no other way are the muscles of the
back and trunk put in perfect repose ; and it must
be readily understood that the more completely
they be rested, the more quickly they recover.   If
a girl be employed at worsted or silk embroidery,
which is now a very common accomplishment for
young people, and has superseded the elegant
occupation of tambour-work, the work-frame
should always be placed at such height and in

such way that she never stoop over it, as by so doing her breathing will be interfered with, and she will not work long before her chest will begin to ache. This may be easily prevented, by the canvas being so placed that the part to be broidered shall be at such height, that, as the arm rests upon it, the elbow and hand shall be nearly on a level with the shoulder. By following this simple rule, work of this kind may be continued for several hours without any chest-ache, and with comparatively little fatigue. It is only carrying out the principle which every prudent person adopts who has to spend many hours at his writing-table. He seats himself in a low chair at a high table, and, with his hand and arm nearly as high as his shoulder, writes for hours with little fatigue, his chest having full play.

Diet.

The mode in which young people are fed is a most important matter as regards their constitutional powers, on which depend the due development of their moral as well as their bodily faculties. That an individual should have a sound mind in an healthy body, he should be well fed, and neither stinted nor stuffed. Parents too commonly err on one or other of these points. Some children, especially girls, are only half fed, for fear of losing their slim figures, and becoming

gross, as it is called. Other children are permitted to cram themselves with everything which
opportunity offers to lay their hands on. Both
suffer from these faults.

A child's diet should be light and nourishing.
Bread and milk are best for his morning and
evening meal; and, if he be not strong, an egg
may be added to his breakfast with advantage.
He should not have meat alternate days, and be
compelled to check his hunger on the other days
of the week with dough dumplings or hard indi-
gestible potatoes, or other less nourishing vegetables. He ought to have plain butcher's meat,
once daily, in proportion to his age and the quickness of his growth and his activity, and in consequence of the two latter circumstances, one
child will, of necessity, require more food than
another, and therefore it is absurd to determine
one particular quantity for all children of the
same age. The proper guide as to the quantity
of animal food a child should have is, how he is
nourished by it, for under the same circumstances
of growth and activity, one child will require
more, while another needs less. The appetite
should be pretty nearly satisfied before pudding
or tart of any kind is put before them; these
should never be served first, as, if they be agreeable to the palate, the child generally manages
to spoil his meat-appetite, and thereby loses the
most important part of his dinner. In schools,

especially those for boys, it is sometimes the practice to make each child eat a certain quantity of boiled hard dough, commonly known by the name of "dampers," a name such abomination richly merits, for thereby the boy is cheated out of his food for the profit of the unworthy pedagogue, who too often in his prospectus talks of liberal board and education. In choosing schools parents should be careful to inquire what the ordinary diet is, and if they can learn beforehand, or find out afterwards, that this precious stuff forms the prologue to the dinner, they should either not place their child there, or remove him as soon as they find it out.

Vegetables in moderation, such as well-boiled mealy potatoes, not the hard ones, which are about as digestible as bricks and mortar, turnips, or almost any which will cook soft, are in general not objectionable. But if a child be weakly and delicate, the food he takes should be as nourishing and as easily digested as possible, and therefore his dinner should be restricted to meat and bread, without vegetable or pudding, except rice, sago, arrow-root, or any other bread-like grain.

I think it advisable that children should have malt liquor at their meals in ordinary cases,— not strong, but certainly sound, and not hard, or inclining to vinegar, as its tonic property is very beneficial. But delicate children should have ale or good undoctored common porter, which

z

strengthens their digestion, at the same time it promotes their appetite. I prefer malt liquor to wine, as its effect is more permanent, and more nourishment is got out of it, whilst wine stimulates only for a short time, and affords no nourishment.

BATHING.

MOST people are aware of the old adage " Clean liness is next to godliness," but comparatively few persons, and young folks especially who have just escaped from the washings and scrubbings of the nursery, attend to it as carefully as its importance requires. A hasty wipe of the face, " a lick and a promise," as it is often not inaptly called, and a careless rinsing of the hands, is too frequently all the cleanliness which boys at least, think necessary, though far from sufficient even if this washing be well performed

In summer time or warm weather washing the whole body by bathing in a stream or in a plunging bath should be regularly practised twice or thrice a week or oftener. But if this cannot be conveniently enjoyed then a shower-bath immediately on getting out of bed may supply its place with great advantage.

The old fashioned shower-bath is a cumbrous and expensive piece of furniture, but in the present cheap and utilitarian times forty shillings will furnish a very useful and convenient apparatus, which has the advantage of occupying little room in the bedchamber. And a still cheaper substitute may be obtained for eight or ten shillings. But should even this sum be of consequence, a very good makeshift is ready at hand in almost every house in shape of a colander. This, however, requires an assistant, who must hold it over the head of the bather with one hand, whilst with the other a jug of cold water is poured through it.

Bathing must not be entered on or persisted in carelessly or it may do great mischief. If the bather do not feel a glow or genial warmth after leaving the water, or when in a shower-bath the water has ceased to fall ; or if he have headache or feel languid and weary, or feel chilly, after the use of either ; then he may be sure the bathing does not agree with him, and he must not go on with it. Care should also be taken as regards the temperature of the water in which the person bathes. Very cold spring water is unfitting for the purpose, as the shock it gives the constitution is very severe. A running stream which by its exposure to the air acquires much of its warmth is best and affords a refreshing cooling. If a shower-bath be used, and first in winter or early spring, it is generally necessary to add, at the onset, a little hot water to take off the chill, and

z 2

this may be gradually lessened in quantity as the constitution becomes accustomed to the shock.

When neither plunging nor shower bath can be conveniently employed, then immediately on getting out of bed sponging the body, the chest more especially, and the arms, is a very good substitute.

Whatever the kind of bathing may be, it is indispensably necessary to rub dry with a hard towel till the person feel a gentle glow, which may be increased by a good polishing with a brush as hard as a horse-brush, which will soon not only be borne without inconvenience, but will afford actual enjoyment.

A *Warm Bath* is a cheap and positive luxury, and is perhaps the readiest restorative, after violent exertion, which can be employed. The heat of the water should not be above 90° or 96° of Fahrenheit's scale, though many accustomed to warm bathing will bear it at 100° or even 110°, but this is too high. Neither should the bather remain in the water above ten minutes, for though at first the sensation be very agreeable to his feelings, yet if the bath be longer continued it may produce faintness and even more serious consequences. It must also be remembered that on leaving the warm bath the person should be careful not to expose himself to a current of cold air, which is very likely by checking the action of the perspiratory vessels of the skin to drive the blood in undue quantity into the vessels of the internal organs and so excite them to inflammation.

WHAT MUST BE DONE IN REGARD TO INFECTIOUS FEVERS.

THERE are some few diseases, to which we are liable in this country, of so infectious character, that when they have once existed in a house there is the greatest difficulty in freeing it of them, and rendering it safe to the present or future residents. Of these, typhus fever, small-pox, and scarlet fever are the most dreaded.

Typhus fever, though generally considered contagious, depends rather on insufficient feeding, bad drainage, crowded dwellings imperfectly ventilated and often surrounded with sweltering filth of all kinds, than on actual contact. Numbers of persons in the same district, and under like circumstances, are attacked with this disease under peculiar conditions of the atmosphere, and hence the disease is considered "catching," as it is called; which, although it must be admitted in very rare instances it is by persons coming from healthy districts, it may be only for an hour or two, into one of these hotbeds of fever, yet, generally speaking, it is untrue; for if the person suffering under typhus fever of the worst form be removed to the airy wards of a general or of a fever hospital, where no precau-

tions beyond the ordinary ones of ventilation and
cleanliness are resorted to—although he may
die, yet very rarely is the disease communicated
to any of the hospital attendants, or to the
patients lying on either side of his bed.

Small-pox and scarlet fever, on the contrary,
are most undoubtedly infectious. The air of the
rooms in which persons subject to these diseases
are lodged, becomes infected by the poisonous
exhalations from their bodies, and is rendered
still more pernicious in proportion to the close-
ness and heat of the chamber; so that seemingly
healthy persons, who may however be in that
state of constitution which renders them favourable to receive an infectious disease, are attacked
soon after they have been exposed to the air
tainted by the disorder. All persons, however,
are not alike susceptible of infection, nor the
same person at all times, for a mother or a
nurse may at one time attend a child with severe
small-pox or scarlet fever, and never take it ; but
on another occasion may nurse a much lighter
case and then be seized by the disorder, and
even destroyed by it. There are few persons
having families who have been so careless as not
to notice this, which is an almost daily occurrence, and therefore shows that the individual
must be in a condition specially suitable for receiving infection.

Now, what is to prevent the continual spring-

ing up of typhus fever, and to prevent the spreading by infection of small-pox and scarlet fever,
or of any other infectious disease ?—for there are
many of them, although I have only selected
small-pox and scarlet fever as being the most
serious.

As regards *typhus fever* the person should be
removed from the filthy hole in which he is too
commonly lodged, if poor, and conveyed to some
more airy and healthy apartment and district.
Towns and parishes ought to be required by
law to have health-houses in proper districts,
independent of the poorhouse infirmaries, whither
the industrious and hardworking, though poor,
artizan—who would die of the disgrace of going
into a poorhouse, and to whom it is the most vil-
lainous injustice that his few traps should be
taken to fit him for the tender mercies of poor-law
relief—might be lodged till his recovery.   Those
who are capable might pay something towards
the expense of such establishments, but others
whose earnings are so little, or their families
so large, as to consume all they earn, and
compel them to live from hand to mouth without
the possibility of laying by anything against
the evil day, should be provided without charge.
If half the money collected for dispensaries,
especially in towns, were employed for the support of health-houses, infinitely more good would
be done to the deserving and needy poor than by

dispensaries, a large portion of the subscribers
to which use them as the means of cheaply doctoring their servants, whose labours they exact
to the utmost whilst they have health, and therefore ought to pay for medical attendance when
they are laid up by sickness in their service,
instead of taking out their yearly guinea sub-
scription's worth from the dispensary, or com-
pelling the servant, if permitted to remain in
their house, to pay the medical man he may call
in from his scanty wages. The latter practice
is of common occurrence, as almost every general
practitioner, except the very disreputable few
who profess not to attend servants, will testify,
even among those who love the uppermost seats
in the synagogue, and are great professors of
charity, so long as their name can be glorified
by the appearance of their guineas in the sub-
scription lists of charities, and which as often
enables the perpetration of jobbery, in a small
way it is true, as any other precious speculation.

If more than one person become sick from
fever in the same family, they should be in like
manner removed to some airy place, otherwise
the longer they continue together, however attentive the medical man may be, and however
largely kind-hearted persons may supply the
needful food, the worse form does the disease
assume, and the more persons in the same room
or house are liable to be attacked with it.

After the removal of the sick, the room should be carefully cleansed, thoroughly washed and dried, as quickly and perfectly as possible, and with as few persons in it as can be arranged. Or, what is still better, it should be completely emptied, and the walls and ceiling lime-washed, and solution of chlorinated lime thrown over the floor. All the small stock of linen should be washed. This may be done at an expense of a shilling or eighteen pence at farthest. But all will be of little service unless provision be made that in future the apartment should be properly ventilated, of which I shall presently speak. This, however, will mend matters only in a degree, nor can we expect that low fever will ever be at certain times less fre-quent in those parts of towns where the poor labouring classes can alone find crowded lodging, till the heartless owners of three and five shilling a week tenements are compelled to lay out some of their exorbitant profits in the proper drainage and supply of water to the alleys and lanes from which they screw their dirty gains. That this at last seems likely to be done, every kind-hearted person who knows in what filth and danger to their health the poor in towns, and often even in large country villages, are compelled to live, will rejoice, and, as far as in him lies, encourage and assist both the Government and associations which are endeavouring to remedy this crying evil, or which, indeed, till within the last few years, but few, excepting medical men and parish officials, were aware.

As regards *small-pox*, the patient should be placed in an airy room apart from the rest of the family, or if the house or cottage will not admit this, he should be removed to some other place. It is always prudent that all the other persons in the house should *at once* be vaccinated if they have not been so previously, or re-vaccinated if they have ; more especially if the person attacked had been vaccinated before taking the small-pox, which now and then occurs. If the vaccination take, it is proof that it was proper it should have been performed, and the party is put in the safest condition possible. But if it do not take, there is no harm done, and the person may consider himself as safe as he can be. From inattention to this advice originates the fatal·spreading of small-pox in certain districts, of which even now we occasionally hear the sad accounts. After the person has left his room, he should, if the weather permit, be as much as possible in the open air, away from other persons, so that his clothes may be purified from the infection of the disease. The apartment should be subjected to the same cleansing as already mentioned, and the house generally as thoroughly cleared as possible of any remaining tainted air, by the windows being kept open in such way as to encourage a free current of air. By attention to these hints all danger of recurrence of the disease usually ceases.

Scarlet fever is, however, the most difficult of all infectious diseases to be got rid of when it has once effected a lodgment in a house. I have known, as there are few observant persons who have not, instances where a house has been subjected to all the ordinary methods of purification two or three times, and yet the disease reappear. I have known an instance in which a school was thrice broken up by three several attacks of this terrible disease immediately after the meeting of the children, although the most careful cleansing of the house, by whitewashing, painting, and the like, had been effected in the intervals ; and I have known where in the same village this disease was continually breaking out in one or other of two large schools alternately, or nearly so, every six months, and for several years together, in spite of all that was done to prevent it.

When a child is attacked with scarlet fever, it should be most carefully separated from all the others of the same house ; and if in a school, or in a large family, unless there be sufficient opportunity to put him completely away from all the others, and in an airy room with a chimney, at the top of the house, he should at once be removed to some neighbouring airy lodging. It is worse than useless to send the other children

away, distributing them among any kind friends
who are willing to take them, as thereby they
often carry with them the disease into other
families. If it be thought proper to remove
them, they should be removed together into a
lodging, and not allowed to have any visitors
beyond their own immediate family. Should
they while there show signs of the disease, they
should be at once brought home, so that the
parent's attention be not divided.

It is of the utmost importance that the air of
the room in which the patient lies should be kept
as pure as possible by continually changing it.
If the weather allow, the tops of the windows
should be down a little and the door open, so
that the air may draw through every part of it.
In winter this can be only imperfectly effected
by a small fire to produce a draught through the
chimney; but the air in the upper part of the
room cannot then be thoroughly changed. It is
very advisable that the bed-furniture and the
window-curtains, more especially if woollen,
should be removed, as they merely form nests for
lodging the foul air. This is far better than
pinning linen wetted with any of the disinfecting
liquids to the bed-furniture, which is of little
real use.

After the child has recovered, it is always
advisable he should be sent into the country,
where he may be in the air as much as possible,

and his clothes should be kept continually out of doors if the weather permit, so that both he and they be thoroughly aired and disinfected.

When the house has been cleared of those who have had the fever, all the bedding should be taken out, and freely exposed in the open air; and every part of the house thoroughly well washed, and the windows continually kept open, so that the air be changed as completely as possible. Some persons whitewash the rooms, and even paper and paint them.

All this may be done, and yet on the return of the inmates the disease will appear again as in the case I have mentioned, and will continue to recur, until *all the drains and cesspools, especially those connected with the water-closets, be emptied and purified.* This is the *alone remedy* by which there is anything like certainty of rendering the house safe from repeated incursions of this dreadful scourge ; and therefore attention to it cannot be too strongly urged.

OBSERVATIONS ON VENTILATION.

No person, who gives the subject a moment's thought, can doubt the necessity of the air in houses being rendered as pure as possible, for the purpose of preserving health; yet how few take the proper means, and how many never make an effort towards this desirable object! It is not merely in the small and crowded tenements of the poor, but also in the more spacious houses of the middling classes, that the same carelessness to ventilation prevails. Nor is it unfrequent that country residences, with every advantage they enjoy, are as badly aired and as close as those in towns, where opportunity for ventilation is far less ready. Many people have so great dread of draughts and catching cold, that they will scarcely ever have a window open; and some, indeed, seal themselves up in their own special atmosphere by having double windows. The only change of air, therefore, which the house enjoys is from that which finds its way in between the house-doors and their frames, and the window-sashes and their frames, and makes its way out as well as it can by the chimney-flues if they be not closed by the registers of the stoves, or by a chimney-board, or by a bag filled with hay, and stuffed into the throat of the chimney. The consequence is that on entering such house, the air feels oppressive, smells frowzy and offensive, and sometimes almost turns one sick. Other persons, though by much the smaller number, run into the very opposite extreme, and every door and window which can be, is put open, and the house blown through and through by draughts of air in every direction. This, indeed, keeps the air pure and fresh, but it is dangerous to the health on account of the currents of cold air to which the inmates are continually exposed.

The object in ventilation is to obtain a constant change of air in a house, without producing violent draughts, which, on account of the general construction, is a matter of great difficulty. The entrance passage or hall, and the staircase, form the principal trunk by which the air from without is distributed to the several chambers, and mixes with that already there, whence a certain portion escapes by the chimneys, in such rooms as have them, and unclosed; but into those without them, the fresh air, speaking widely, never enters, and if it do manage to steal in, cannot get out, consequently the air there remains almost completely unchanged—a state of things not at all uncommon in the attics of small and even middling-sized houses, in which builders too commonly have a most unaccountable propensity not to make a chimney-flue. Another very objectionable point is that the heads of the windows rarely

reach to the ceiling, but more commonly in lofty
rooms not within two or three feet of it, and in
those of more moderate proportions not within
twelve or eighteen inches, the consequence of
which is, that although every door and window
be set wide open, all the air above the windows
is but very imperfectly changed.

To secure even a certain amount of ventilation, it is absolutely necessary that every room
should have either a fireplace or a flue by which
the air may escape, and thereby a current
produced. Windows alone are insufficient for
this purpose, as although in fine and warm
weather they may be kept open throughout the
day, yet in wet weather and during the night this
cannot be permitted; and therefore for many
hours, and at night especially, when the room is
closely shut, and one or more persons sleeping in
it, the air is quickly and greatly fouled, and very
injurious to the health of the sleepers. Also, if
persons be confined by sickness day and night in
such rooms, the air is still further spoiled; and
should they be suffering from any infectious disease, the danger, to themselves and those wh ;
attend them, is greatly increased by the air becoming more and more foul and infectious.

Even though the room have a chimney, the
air in it is only partially changed, for a large
quantity in the upper part of the room will not
descend and pass away by the chimney: it will

lurk about the ceiling and in all the corners, and
thence can only be made to move by passing
through the windows when their upper sashes
are dropped ; and only perfectly then, if the
window-heads reach quite to the ceiling. It is
for this reason that the windows of all rooms,
and specially of sleeping-rooms, should reach the
ceiling, as the only means by which the air can
be completely changed. Every one must be
well aware that when a fire is burning in the
stove, a very large portion of the air which enters
beneath the door makes its way directly to the
fireplace ; and hence it is, that though the upper
part of our body is pleasantly, and sometimes
more than pleasantly, warmed by the heat which
the fire throws out, the current of cold air running
along the bottom of the room is so brisk, that we
feel as if our legs were in a pail of cold water,
or, as it is not inaptly expressed, we feel as if
our legs are almost cut off with the cold.

To prevent this, mats are put close to the
door, and its edge is either leathered or listed ;
and to prevent the air drawing in between the
sashes and frames of the windows, the curtains
are closely drawn. This sometimes, though not
always, keeps the cold air out, or admits so little
that we are not inconvenienced by it. But we
are soon plagued in the contrary way by the
room becoming so hot that one might as well sit
in an oven, and are therefore compelled to open

the door, and admit a large quantity of air, which suddenly cools the room ; after which the door is closed, the room becomes heated again, and the same proceeding is necessarily repeated several times during the evening, if the apartment have many persons in it, or if it be very small, and is as inconvenient as it is unhealthy on account of the constant variation of the temperature.

Now any means by which the air coming in from the door and windows can be prevented at once finding its way to the fireplace, and made to spread itself about the room, and mix with the air already there, so that it may generally change and refresh it before it pass away, will check the constant current of cold air along the floor, and prevent the air in the upper part of the room remaining unchanged and foul when the room is closed, especially for the night.

A vast number of expedients have been proposed for this purpose, too numerous to be noticed here. The object is, or should be, to make an escape for the heated foul air at the upper part of the room, in such way that no violent draught should be produced, whilst a circulation of air is at the same time produced in every part of the chamber. The best mode of doing this would be that all ceilings should cove towards the centre, in which should be an aperture leading to a flue carried between the ceiling and the flooring above, and thence running up within the wall, or within the rooms above, to the top of the house, either within or without the roof, and there opening, but provided with a whirling fan, so arranged that while it allows the escape of the air from below, it prevents the descent of cold air from above. In this way, without difficulty, the air in the upper part of the room is changed, and that heated by its nearness to the fire, rising by its lightness, becomes mixed, renders the temperature throughout pretty equal in a room of moderate proportions, and assists in the circulation of the fresh air, without draughts of any consequence being produced.

In rooms with flat ceilings nearly the same results may be produced by cutting a hole or holes, according to the size of the room, through the ceiling, and making the space between any two joists above a horizontal flue, which by another hole or holes in the floor above may be made to communicate with a wooden vertical flue placed against the wall, and running up through the upper rooms till it reach the roof. This requires no great stretch of intelligence nor much expense to carry into effect, and might be at any time, and under almost any circumstances, whether the house be large or a mere cottage, arranged without difficulty.

Within the last few years, a very simple and ingenious contrivance has been invented by Dr.

NEIL ARNOTT for ventilating rooms, which very effectually answers the purpose, and has the advantage of being very cheap, so that for a few shillings all the desired advantage may be gained. "The Ventilating Valve," as the Doctor calls it, "is a contrivance placed in an opening made into the chimney-flue near the ceiling (not nearer than nine inches to any timber or other combustible substance, according to the new Building Act), by which all the noxious air is allowed at once, in obedience to the chimney-draught, to pass away, but through which nothing can escape outwards [it should rather have been backwards into the room.—J. F. s.] or return. Contraction of the chimney-throat, by the register of a good open grate or stove, aids its action. It is useful in all cases to have the register door, and when there is no fire in the grate it should be shut. The valve is in principle a small weighbeam or steelyard, carrying on one arm a metallic flap to close the opening, and on the other a weight to balance the flap. The weight may be screwed on its arm to such a distance from the axis or centre of motion, that it shall exactly counterpoise the flap; but when left a little further off, it just lifts the flap very softly to the closing position. Although the valve, therefore, be heavy and durable, a breath of air suffices to move it, which if from without (that is, from within the room) opens it, and if from within (that is, from the chimney-flue) closes it, and when no such force interferes it settles in its closed position." It may be obtained at ED-WARDS's, Poland Street, Oxford Street.

I may here notice that if the weather permit, the window-sashes of all the sitting-rooms should be opened, so that the air may enter, both above and below, for about half an hour so soon as the house is opened in the morning, and those also of the sleeping-rooms, as soon as their occupants have left them, by which time the windows of the lower chambers are probably closed, so that there is no very violent current of air through the house, although there be, in this way, sufficient to air the rooms thoroughly in succession.

Now though the methods which I have just mentioned, and ARNOTT's especially, will certainly very effectually discharge the foul air from the room, yet I am aware that it will not do away with entirely, although it will diminish, the rush of the cold air along the floor to the fire-place. To get rid of this very great nuisance, Professor HOSKING* recommends "that a sweet-air flue should be made within the outermost jamb of

* Health of Towns Commission, First Report, 1843.
In the diagram, A A mark the air flues; and the arrows show the direction of the current of air.

every chimney breast, from the bottom to the top, and opening into the outer air free from communication with the smoke-flues, and not liable to be contaminated by them. This should open at the lowest level again, either directly under the floor, or by a horizontal flue with a grated mouth, so that it may be fed with fresh air at both ends. Every fire should be fed, by an opening from such flue behind the cheeks and back of the grate, with the fresh air ; perforations underneath the fire-grate admitting it to the fire, and other perforations in the cheeks admitting it to the room generally ; so that the air passing round and about the back of the grate would receive heat from the substance of the grate and enter the room warm." This method of warming the air before its admission into the room, Professor Hosking has employed with very excellent effect ; and, in connexion with Dr. Arnott's ventilating-valve, forms a complete system, applicable to all rooms in which there are chimneys, and no room can fail to be wholesomely ventilated in which it is properly carried out.

POISONS

BESIDES those cases in which poison has been intentionally taken or administered, it not unfre-quently happens that persons are accidentally poisoned. Thus, oxalic acid has sometimes been swallowed by mistake for Epsom salts; an in-stance is recorded of a girl sixteen months old having died in consequence of putting into her mouth some pieces of "blue-stone" that were given her to play with; and overdoses of lauda-num, or wine of colchicum, have proved fatal.

The first thing to be done in most cases is to get rid of the poison from the stomach, before it enters into the system; and no time ought to be lost in setting about it. Nature often performs this office to some extent itself, for vomiting is a very usual accompaniment of improper substances being taken into the stomach: but, if this action do not already exist, it must be produced. About half a drachm, that is a quarter of a teaspoonful of sulphate of zinc ("white vitriol") dissolved in a large wine glass full of warm water will almost always effect this; and the dose may be repeated in a quarter of an hour, if it appear desirable. Ipe-cacuanha, antimonial wine, or any other emetic

may of course be employed ; but if these be not
at hand, one or two teaspoonsful of powdered
mustard mixed up in a glass of water, and taken
at intervals, will be found a good substitute.
The retching should be kept up by swallowing
soapy liquids, linseed tea, milk, or warm water ;
indeed, it is always important that such drinks
should be freely made use of, in order to catch
hold of the poison, and cause it to be brought away
by the vomiting, especially if it have been taken
upon an empty stomach.   A repetition of gentle
blows upon the back between the shoulders, or
tickling the throat, may be resorted to as provo-
cative of retching.   If a surgeon arrive, he will
apply the stomach pump.

Although it may be taken as a general rule
that the poisonous substance ought to be got rid
of by vomiting or purging, it is not a rule without exceptions ; and in any case it is only the
first step : the subsequent treatment depends
entirely upon the nature of the poison, and must
be described accordingly.   And, even though the
case has been so successfully treated that no
danger is to be apprehended from the immediate
effects of the poison, yet it is better to send for a
medical man, if possible, as some of the symptoms
which may ensue long afterwards, may prove
serious.

Sulphuric acid, or Oil of Vitriol—Nitric acid, or Aqua Fortis.

These are among the cases in which the attempt to produce vomiting will be of little service, for the acids in question act instantaneously upon all the membranes of the mouth, throat, and stomach, decomposing and dissolving them : indeed, the very act of retching would tend to bring away blackened pieces of the inner lining of these organs of the poor sufferer. It is much better to lose no time in neutralising the acid. This may be done by swallowing soda or potash dissolved in a considerable quantity of water. Magnesia, or carbonate of magnesia, may be taken instead of this, and milk is a good vehicle to take it in. Or a dose of " Dinneford's solution " may be drunk just as it is. In the absence of any of these substances, whiting, lime, or powdered chalk, mixed up with some innocent liquid, should be taken : or some of the whitewash may be scraped off the wall or ceiling, and swallowed by the unfortunate patient along with a quantity of water.

Oxalic acid.

Vomiting generally takes place naturally after oxalic acid has been swallowed. The proper antidotes are, magnesia, carbonate of magnesia, lime, whiting, or chalk, mixed up with water, milk,

gruel, or any other convenient liquid. These generally prove efficacious if administered speedily.

Arsenic

Is a well-known deadly poison, and unhappily often
resorted to by the murderer and suicide. It is a
common agent in secret or slow poisoning; but
fortunately the advance of science has rendered
the detection of this, and most other noxious substances in the body of the deceased, so easy, that
no such wholesale tragedies as those perpetrated
by means of the Aqua Tolfana, or the Eau de
Brinvilliers, could now remain long undiscovered.

The general symptoms of poisoning by arsenic
are—a feeling of sickness and faintness commenc-
ing half an hour after it has been taken; great
tenderness and pain in the stomach, succeeded
presently by violent retching attended with thirst.
Sometimes diarrhœa and tightness of the chest
accompany these symptoms. A general feeling
of illness and pain then takes possession of all
parts of the body, and death ensues in a period
varying from twenty-four hours to three days.

In whatever form arsenic may have been taken
emetics and purges are most to be relied on for
saving the person's life: yet violent emetics should
be avoided, if sickness can possibly be produced
by milder means, as they tend to increase the
irritation that already exists on the coats of the
stomach. Small quantities of milk given at inter-

vals will facilitate the vomiting; and the white, or glare, of raw eggs may be swallowed—it will at least be innocent, and may prove highly serviceable. Professor Orfila speaks of " a quack, on whom twelve grains of arsenious acid ('white arsenic') produced little or no inconvenience, because he drank before it a great quantity of milk, which was quickly vomited, together with the poison." If the materials are comeatable, take a few crystals of " green copperas ;" dissolve them in warm water ; add several drops of nitric acid (" aqua fortis ") ; add common soda, stirring it well, until the mixture ceases to taste inky. Let the patient swallow the thick reddish brown draught, that will thus be prepared—it is a mixture of the aperient " Glauber's salts," with the best antidote yet discovered for arsenic.

Corrosive Sublimate.

It is impossible that any one should take a quantity of this substance without being instantly aware of it from the excessively disagreeable coppery taste which it has. The white, or yolk of raw eggs is the proper antidote for this poison; but, if it should so happen that no uncooked eggs are in the house, flour made into a paste with water will serve the same purpose. Yet it will be advisable in every instance to send for some eggs, and as soon as they arrive beat up the whole contents of the shell in cold water, and drink it. The egg, or flour, too, as well as being an antidote, will act beneficially in increasing the vomiting; and this must always be well attended to.

Sugar of lead

Has much the appearance of crystallized white sugar, and resembles it in taste. A small dose does not seem to produce any more serious result than a stomach-ache; a large dose may prove fatal. In the treatment of persons poisoned by this substance we are fortunately situated, for the very medicines which are most suitable for getting rid of the poison render it harmless too. Sulphate of zinc, whilst it acts as an emetic, serves also as an antidote to lead; and Epsom salts or Glauber's salts are also antidotes, whilst at the very same time their purgative qualities assist in carrying off the noxious matter.

Slow poisoning by lead is very common. It is the "lead colic" of painters. Pure water kept in a leaden cistern, or flowing through a considerable length of leaden pipe, dissolves a small quantity of the metal, and those who habitually drink, or cook their food in such water, are slowly affected by it. Constant gripings in the stomach and bowels are the first symptoms of this sort of poisoning, and they are succeeded by palsy, the hands hanging down as though the wrists had no strength to support them. A blue line on the teeth along the edge of the gums is also a cha-

racteristic indication. If attacked by these disagreeable symptoms, the sufferer should of course
consult a doctor: no immediate relief can be
obtained, unless what warm baths and opening
medicine may afford. But " prevention is better
than cure ;" hence it is recommended that every
one should avoid as much as possible all connexion with lead in any form. But as there are
very many, who, from the nature of their business
or other circumstances, cannot act upon this principle, they should be particularly attentive to
cleanliness ; changing their clothes after the day's
work is done, and taking especial care always to
wash their hands and lips before eating. A
minute quantity of sulphuric acid put daily into
the beer, or other drinks, of workmen exposed to
danger from preparations of lead, has been found
beneficial. It is astonishing how unsuspected
may be the cause which has brought about this
slow poisoning. A woman was lately admitted
into one of the London hospitals with pain in the
stomach, palsy, blue line along the gum, and, in
short, every symptom of lead colic ; but she was
the wife of an upholsterer, and could not conceive
in what manner she had been subjected to the
influence of lead, until it was recollected, a good
while afterwards, that some of her sons had
amused themselves the preceding summer by
making birdcages, and painting them green in
the room in which all the family lived. The

green paint was at once an explanation of the mystery.

Soda, or Potash.

If any one should happen to swallow a quantity of dissolved pearlash, potashes, soap-lees, soda, or the carbonates sold by druggists, he will immediately become aware of the fact by their acrid caustic taste, and by the burning heat in his throat, which will directly ensue. The proper course to be pursued in such a case is, to drink the first innocent acid that comes to hand mixed with a quantity of water. Vinegar, or lemon-juice, will answer as well as anything. Milk, gruel, or barley-water may be drunk, and sour fruits may be eaten with advantage. Large doses of oil have also been strongly recommended.

Prussic Acid.

This is the most terrible of the poisons, on account of the rapidity with which it destroys life. It is not even necessary that any of the liquid itself be swallowed, for the smell alone is sufficient to put an end to mortal existence. I know a gentleman who once, out of curiosity, put a bottle of prussic acid to his nose in a chemical store-room: almost immediately he felt (as he described it) all his blood fly to the extremities of his limbs, and his jaw become spasmodically fixed, while he nearly fell on to the floor. He had however

enough strength and presence of mind to snatch
a bottle of ammonia, breathe some of the gas,
and rush into the open air. A few minutes
afterwards he returned, and quietly told us what
had happened. He walked about the streets for
the remainder of the day, as he felt very queer
whenever he rested ; and it was some days before he completely recovered. As this acid is
constantly given off from the cyanide of potas-
sium employed now in some of the arts, it is as
well to be on the guard against too strong a smell
of it. The odour, too, acts very differently upon
different individuals : indeed the men employed
in electro-plating will work comfortably in an
atmosphere so charged with prussic acid, that it
will instantly give a headache to others who have
not been acclimatized to it.

If you should find yourself in a position similar
to that of the gentleman just referred to, you
cannot do better than he did : you should breathe
as much as possible of the pungent fumes of *liquor
ammoniæ* (spirits of hartshorn), or *sal volatile*, and
you should dash cold water over your person.

If you should find any one who, designedly or
otherwise, has just taken prussic acid, you may
recognise the fact by a peculiar odour of bitter
almonds there will be in the apartment, by the
glistening and dilated appearance of his eyes, by
his insensibility, and by his irregular breathing,
sometimes gasping convulsively, and then again

sobbing. As it is then utterly impossible he can help himself, and he may die in a minute or two, you must not lose a moment in giving him the restoratives above mentioned. A common smelling bottle may save life. You should not fail also to pour cold water from a height upon his back, or between the shoulders. If any bleaching powder, or chloride of lime, be near at hand, the fumes given off from it may prove very serviceable; and in order to produce more of the gas, a little vinegar may be poured upon the stuff, taking care, though, not to suffocate the patient in your haste to restore him.

There are several things in common use which contain a small amount of prussic acid, and are on that account poisonous if taken. Oil of bitter almonds is one of these; and bitter-almond water, laurel water, and cherry-laurel oil are of the same character. They should, of course, be guarded against.

Opium; Laudanum.

More persons are destroyed by opium and its various preparations, than by any other poison. Besides the substances placed at the head of this division, there are a number of others containing it, which are constantly sold at the apothecary's :— Dover's powders, syrup of poppies, Godfrey's cordial, &c.; and it is lamentable to think how many infants are brought to an untimely grave because their mothers, to save a little trouble, will drug them with such opiates. Muriate, acetate, and sulphate of morphia also contain the same principle in a still more potent condition.

The quantity of opium necessary to destroy life appears very various. A case is recorded, in which a powder, containing only two-fifths of a grain of it, killed a child four years and a half old; whilst, on the other hand, some people have accustomed themselves to take large doses of laudanum without injury; and the quantity consumed by a confirmed opium-eater is notorious. In one respect a large dose may prove less dangerous than a comparatively small one, as it frequently causes vomiting, and thus ejects itself before it has had time to do much mischief.

In young children the effects of the poison will be almost immediately apparent, but in the case of adults half an hour, or an hour, usually elapses. The attack begins with giddiness; this is succeeded by determined drowsiness and stupor; the pulse is quick and irregular, the breathing hurried, and there is copious perspiration. In fatal cases death commonly ensues in from six to twelve hours.

The smell of opium is peculiar, and often leads to a knowledge of the real state of the case.

Emetics should be administered if the poisoned party can in any way be made to swallow. Strong

2 B

coffee should also be given ; and cold water is to
be poured upon the head, chest, and back. In
the case of little children it has been found advantageous to plunge them into a warm bath,
and then suddenly carry them into the cold air;
but the most important matter to be attended to
is to keep the patient awake. He will show every
disposition to yield to the drowsiness which op-
presses him, and will perhaps remonstrate with
those who would prevent his sleeping the sleep of
death ; but in spite of that he should be walked
up and down the room between two strong men ;
pulling the hair, or any other device, which in-
genuity may suggest for rousing him, should be
resorted to ; and he should be allowed no rest
until all danger is over.

Poisonous Mushrooms,

Or "toadstools," as they are familiarly called,
often produce serious accidents. Perhaps five or
seven hours after the unwary individual has par-
taken of them, he is attacked with gripings in the
stomach, disorder in the bowels, and afterwards
cramps, convulsions, and unconquerable thirst;
or sometimes giddiness in the head, with a stupid
delirium or drowsiness. The symptoms, however,
vary much according to circumstances, and the
kind of mushroom eaten. Christison quotes the
following narrative :—" A man gathered, in Hyde
Park, a considerable number of the *Agaricus*

campanulatus, which he mistook for the *A. campestris*, stewed them, and proceeded to eat them; but before he had concluded his repast, and not above ten minutes after he began it, he was suddenly attacked with dimness of vision, giddiness, debility, trembling, and loss of recollection. In a short time he recovered so far as to be able to go in search of assistance. But he had hardly walked 250 yards when his memory again failed him, and he lost his way. His countenance expressed anxiety, he reeled about, and could hardly articulate. The pulse was slow and feeble. He soon became so drowsy that he could be kept awake only by constant dragging. Vomiting was then produced by means of sulphate of zinc; the drowsiness gradually went off; and next day he complained merely of languor and weakness."

No other general advice can be given in cases of this kind of poisoning, beyond the use of emetics. Pieces of the mushroom are often brought up undigested.

Poisonous Mussels, or other food.

It often happens that people are attacked with alarming symptoms after eating certain kinds of fish, especially mussels, for no reason that can be assigned. Meat beginning to putrefy, and the flesh of cattle which have been overdriven, or have died of disease, may communicate a virulent poison to those who make use of such food.

Among poisonous plants may be mentioned
henbane, deadly nightshade, tobacco, hemlock,
monkshood, foxglove, and nux vomica. Among
the principles to which these owe their poisonous
qualities are, strychnia, belladonna, and veratria.
The powerful properties of these are sometimes
taken advantage of by the medical practitioner;
but if they are incautiously taken inwardly, they
prove fearfully destructive to human life. No
antidote is known for any of these. Orfila recommends that after emetics and purges, bleeding
should be resorted to, and acidulated drinks
should be administered; and subsequently, attention must be paid to keeping down the inflammation, that will in all probability ensue.

THE END

INDEX.

2 c

2 c 2

LONDON: PRINTED BY WILLIAM CLOWES AND SONS, LIMITED
STAMFORD STREET AND CHARING CROSS